Manual of Water Supply Practices

M66

Cylinder and Vane Actuators and Controls—
Design and Installation

American Water Works Association

M66

Cylinder and Vane Actuators and Controls—Design and Installation

American Water Works Association

Manual of Water Supply Practices—M66

Cylinder and Vane Actuators and Controls—Design and Installation

Disclaimer
The authors, contributors, editors, and publisher do not assume responsibility for the validity of the document or any consequences of its use. In no event will AWWA be liable for direct, indirect, special, incidental, or consequential damages arising out of the use of information presented in this book. In particular, AWWA will not be responsible for any costs, including, but not limited to, those incurred as a result of lost revenue. In no event shall AWWA's liability exceed the amount paid for the purchase of this book.

Sr. Manager, Editorial Development and Production: Gay Porter DeNileon
Sr. Production Editor: Sandra Lankenau

If you find errors in this manual, please email books@awwa.org. Possible errata will be posted at www.awwa.org/resources-tools/resource.development.groups/manuals-program.aspx.

Library of Congress Cataloging-in-Publication Data

Cylinder and vane actuators and controls: design and installation. — First edition.
pages cm — (AWWA manual; M66)
Includes bibliographical references and index.
ISBN 978-1-62576-116-3 (alk. paper)
1. Slide-valves. 2. Hydraulic servomechanisms. 3. Pneumatic control valves. 4. Blades. 5. Waterworks. I. DeNileon, Gay Porter, editor. II. American Water Works Association.
TD491.C95 2015
628.1028'4--dc23

2015035968

ISBN-13 978-1-62576-116-3
eISBN-13 978-1-61300-347-3

Printed in the United States of America
American Water Works Association
6666 West Quincy Avenue
Denver, CO 80235-3098
awwa.org

Printed on recycled paper

Contents

Figures

Tables

This page intentionally blank.

Preface

This new AWWA Manual of Water Supply Practices presents the general principles of hydraulic and pneumatic power actuators for use with waterworks valves. The principles used for selecting, sizing, and installing actuators are detailed. This is a discussion of recommended practice, providing guidance on generally available actuators and their application, not an American Water Works Association (AWWA) standard. Questions about specific situations or applicability of the guidance provided here to specific valves and actuators should be directed to the manufacturer or supplier.

Operators, technicians, and engineers will find the information contained in this manual useful for understanding the use and application of hydraulic and pneumatic power actuators. There are many special actuators that are beyond the scope of this manual and require special consultations with those suppliers. The valve thrust and torque information is generic, and actual data from the valve manufacturer should be obtained before making the final selection of actuator size and options. This manual provides information on the hydraulic and pneumatic actuators listed in AWWA Standard C541 Hydraulic and Pneumatic Cylinder and Vane-Type Actuators for Valves and Slide Gates, latest edition, including the following:

- Linear cylinder actuator
- Quarter-turn cylinder actuator
- Quarter-turn vane actuator

This manual refers to AWWA standards, which are available for purchase from AWWA by calling (800) 926-7337 or online at www.awwa.org.

Manufacturers graciously provided valve illustrations and other documentation. AWWA does not endorse any manufacturer's products, and the names of the manufacturers have been removed from the material provided.

Metrification note: Cylinder sizes are listed in their current US designation: inches. To obtain an approximate metric equivalent, use a conversion factor of 25.4 mm per inch.

This page intentionally blank.

Acknowledgments

The AWWA Subcommittee on AWWA Manual M66, which developed this manual, had the following personnel at the time of approval.

John V. Ballun, *Chair*

J.V. Ballun, Val-Matic Valve & Manufacturing Corporation, Elmhurst, Ill.	(AWWA)
B.E. Bosserman, Engineering Consultant, Mission Viejo, Calif.	(AWWA)
G. Eaton, Festo Corporation, Dallas, Texas	(AWWA)
M. English, Pentair Valves & Controls, Pelham, Ala.	(AWWA)
F.L. Hinker, Consulting Engineer, Santa Rosa, N.M.	(AWWA)
J.R. Holstrom, Val-Matic Valve & Manufacturing Corporation, Addison, Ill.	(AWWA)
M. Ford, Kinetrol USA Inc., Wakefield, R.I.	(AWWA)
B. James, TTR Group Inc., Oakville, Ont.	(AWWA)
A.W. Libke, DeZURIK, Sartell, Minn.	(AWWA)
M. MacConnell, Metro Vancouver, Burnaby, B.C.	(AWWA)
W.H. Peffley, Crawford Murphy & Tilly, Springfield, Ill.	(AWWA)
F.R. Underwood, K-Tork Actuators & Controls, Dallas, Texas	(AWWA)
J. Wilber, American AVK, Littleton, Colo.	(AWWA)

The AWWA Standards Committee on Power Actuating Devices for Valves and Slide Gates, which reviewed and approved this manual, had the following personnel at the time of approval:

Fred L. Hinker, *Chair*
Amzad Ali, *Vice-Chair*

General Interest Members

A. Ali, ADA Consulting, Surrey, B.C.	(AWWA)
B.E. Bosserman, Engineering Consultant, Mission Viejo, Calif.	(AWWA)
F.L. Hinker, Consulting Engineer, Santa Rosa, N.M.	(AWWA)
K.H. Hipps, CDM Smith Inc., Chicago, Ill.	(AWWA)
T.J. McCandless[1], Standards Engineer Liaison, AWWA, Denver, Colo.	(AWWA)
J.P. Musich, Wright-Pierce, Topsham, Maine	(NEWWA)
W.H. Peffley, Crawford Murphy & Tilly, Springfield, Ill.	(AWWA)

[1] Liaison, nonvoting

Producer Members

R. Arnold, Rotork Controls Inc., Rochester, N.Y. (MSS)

J.V. Ballun, Val-Matic Valve & Manufacturing Corporation, Elmhurst, Ill. (AWWA)

P.E. Brunelle, Rodney Hunt Company, Orange, Mass. (AWWA)

L.H. Elliott, EIM Company Inc. Valve Controls, Missouri City, Texas (VMA)

J.R. Holstrom, Val-Matic Valve & Manufacturing Corporation, Addison, Ill. (AWWA)

M.D. Rooney[2], Emerson, Lynchburg, Va. (AWWA)

F.R. Underwood, Rotork, Dallas, Texas (AWWA)

User

S. Carpenter, San Diego County Water Authority, Escondido, Calif. (AWWA)

T.W. Knowlton, Salem & Beverly Water Supply Board, Beverly, Mass. (NEWWA)

M. MacConnell, Metro Vancouver, Burnaby, B.C. (AWWA)

P. Ries, Denver Water Department, Denver, Colo. (AWWA)

[2] Alternate

Chapter 1

Introduction

Hydraulic and pneumatic actuators are externally mounted on ball, plug, and butterfly valves, and on slide gates to provide automatic control of these devices in water and wastewater applications. When combined with control devices such as solenoid valves or positioners, the valves and gates can be used to remotely control water and wastewater treatment processes or process variables including flow and pressure. Many of these actuators also have the feature of providing field-adjustable operating times and, when equipped with pneumatic or hydraulic power systems, can continue to operate after loss of electrical power.

HISTORY

Simple gate valves date back to ancient times in Greek and Roman water and irrigation systems. These gates were mostly operated by manual levers or screws. It was not until the industrial revolution and the use of steam power that valves became automated with power actuators. Since then, hydraulic and pneumatic actuators have been a staple in water and wastewater systems, particularly as valves grew in size and became extremely difficult to operate manually. The advent of automation made it desirable to operate valves remotely with computerized process control systems. Figure 1-1 illustrates a ball valve equipped with a pneumatic vane-type actuator that controls the flow from a high-service water pump.

Given the increased use of automatic process systems, the use of power actuators continues to grow. The power actuator positions the valve in response to system-control signals to complete a system-control loop. When flow or pressure must be controlled as part of a process, a sensing device such as a pressure transmitter or flowmeter sends a signal to a process controller, which in turn compares the signal to the desired system parameter and sends a position signal to the power actuator. The power actuator is directly coupled to the closure member of the control valve and is therefore an essential element of the control loop (Skousen 1997).

Figure 1-1 Quarter-turn vane-type actuator provides remote control of a ball valve
Source: Courtesy of Val-Matic Valve & Mfg., Corp.

Since a number of valve types are used to control processes, a similar number of various actuator types are used to control these valves. In response, the valve industry has developed actuators for both linear and rotary valves. When automated valves are needed, either hydraulic actuators or electric motor actuators are specified. Electrical actuators provide great versatility in valve automation, but hydraulic actuators provide the ability to store energy and operate the valve without electrical power. Hydraulic systems also provide safe operation of valves, which may be located in hazardous environments. Furthermore, in the event of an electrical power outage, hydraulic and pneumatic systems can still function to provide safe shutdown or isolation of the system.

Electric and hydraulic actuators are used regularly in many industries, including oil, gas, and power, but their use for the waterworks industry was standardized by AWWA Standard C540 in 1987. In 2008, AWWA C540 was withdrawn and replaced by standards AWWA C541, Hydraulic and Pneumatic Cylinder- and Vane-Type Actuators for Valves and Slide Gates, and AWWA C542, Electric Motor Actuators for Valves and Slide Gates. Note, like all AWWA standards, C541 and C542 are updated regularly, as necessary depending on new technology and requirements.

PURPOSE

The purpose of this manual is to explain the fundamental concepts of hydraulic and pneumatic power actuators so users with a technical background can understand, specify, and install these actuators on valves and slide gates. The information in this manual must be used in conjunction with data from actuator and valve manufacturers to select and size the correct actuator models for specific applications.

Sizing parameters such as thrust and torque are explained so the sizing methodologies recommended by actuator manufacturers can be applied to specific applications. Also, the effects of certain valve application parameters, such as flow rate or differential pressure, on actuator operation are discussed.

Finally, installation and operation guidelines are presented for startup procedures and troubleshooting power actuators in the field.

GENERAL APPLICATION

Cylinder-type actuators produce a linear motion to operate a slide gate. Hydraulic or pneumatic pressure is applied to either side of the cylinder piston to generate linear force or thrust. The cylinder rod is attached to the gate and causes it to rise and fall inside the gate frame.

Quarter-turn cylinder-type actuators produce a rotary motion to turn the valve stem and operate the closure member of the butterfly, plug, or ball valve. Quarter-turn cylinder actuators convert the linear motion of a cylinder to rotary motion using link-and-lever, rack-and-pinion, or scotch-yoke mechanisms. Each of these mechanisms is discussed in chapter 2, "Principles of Operation."

Quarter-turn vane-type actuators also produce a rotary motion. However, instead of a cylinder, the vane acts like a lever and generates rotation from the pressure applied to either side of the vane.

When specifying power actuators, consideration should be given to the effects of valve operational speed on the creation of pipeline hydraulic transients (surges), especially on long pipelines. The power actuator stroke time may vary based on actual pipe operating fluid conditions (e.g., pressure, flow, and pipe size) and the power source capacity.

SUPPLY MEDIUM

All of these actuators are powered with a supply medium of air, water, or hydraulic oil pressure. Specific requirements for the supply media, including pressure, flow, and cleanliness, will be discussed further in this manual. When accessories and control devices such as solenoid valves are added to the actuator, the supply media requirements become more critical.

All three supply media—air, water, and oil—provide reliable sources of energy to operate valves, even after loss of electrical power. The supply systems can be installed near the valves or up to several hundred feet away if required. Both indoor and outdoor installations for these systems are available from either the valve or control system supplier. A comparison of the most relevant properties of these three supply media is summarized in Table 1-1.

Air Supply Medium

Air is a common power supply medium because it is relatively inexpensive and often readily available at water and wastewater plants to operate pneumatic power tools. Compressed air systems consist of an electric motor–driven compressor and a receiver tank for air storage. These systems commonly provide air pressure in the 80–100 psig (552–689 kPa) range with a maximum recommended pressure of 150 psig (1,034 kPa).

The supply output from the compressor system should be equipped with filters and dryers so debris down to a size of 40 microns and water droplets are removed from the air supply before distribution. When the valve is equipped with modulating controls or positioners, filtration down to 5 microns is recommended to prevent clogging of the control orifices. When recommended by the valve or actuator manufacturer, lubricators can be used to add lubrication to the cylinder actuator, but they should not be added to vane actuators or when positioners are employed. It is important to check with the actuator manufacturer first because most vane actuators are permanently lubricated and lubricated air may displace the manufacturer's supplied lubricant from the actuator.

Table 1-1 Summary of media properties

Property	Air	Water	Oil
Relative cost of media equipment	Moderate	Low	High
Pressure ranges	80–100 psig (552–689 kPa)	60–150 psig (414–1,034 kPa)	60–150 psig (414–1,034 kPa) 1,000—2,500 psig (689,–17,237 kPa)
Speed control	Poor (large cylinders)	Good	Good
Required cylinder maintenance	Low	High	Low

While air is an economical and reliable power supply medium, its use should be limited to smaller valves. If used on large valves, the size of the actuator cylinders can become too large to dispense the required actuator force because of the 80–100 psig (550–690 kPa) pressure range limitation. Also, if the valve needs to be moved slowly (i.e., over one minute), the motion of the cylinder may not be smooth. Air is compressible, and the forces from the valve vary with position, which causes the pressures in the cylinder to fluctuate, preventing smooth motion of the cylinder piston. Also, when air is released from the cylinder, it can produce uncomfortable or irritating noise levels if not fitted with proper sound-reducing mufflers.

Water Supply Medium

Water is also a common power supply medium because it is typically available at plants and pumping stations, especially if the plant or pumping station is feeding the water distribution system. A reliable water supply for powering actuators can often be obtained by connecting to a pump discharge header . However, it is important to realize that during system startup, the water source may not yet be available, and a temporary water supply source may be needed to operate the valves. If potable water is used as the supply medium, care must be taken to prevent backflow of the supply and drainage water back into the potable water system. Depending on the source of the water, providing filtration to remove particles larger than 20 microns may also be necessary. Other properties of the water, such as salinity, hardness, pH, and conductivity, should be reviewed to protect the cylinder and controls from corrosion, degradation, and excessive maintenance. A minimum water pressure of 60 psig (414 kPa) is common and a maximum pressure of 150 psig (1,034 kPa) should be used to match the standard pressure rating of the cylinders.

Water is not compressible and therefore can provide smooth operation of cylinders at extremely slow operating times, such as five minutes. It may be necessary to open or close a valve this slowly to prevent surges in long distribution pipelines. Water in general is corrosive, so its use dictates the need for corrosion-resistant supply tubing and cylinders. Water supply tubing to the actuator is usually brass and copper. Similarly, the cylinder barrel should resist corrosion and be constructed of materials such as stainless steel, bronze, fiberglass, or coated steel. The cylinder rod should be of corrosion-resistant material as well, such as hard-chrome or electroless-nickel-plated stainless steel. Water-powered actuators are not generally certified to drinking water standards such as NSF 61,

Drinking Water System Components – Health Effects, so the exhaust media from the actuator should not be returned to the potable water source.

Hydraulic-Oil Supply Medium

The hydraulic-oil supply medium is typically a petroleum-based fluid, but it can also be certified as food grade per FDA requirements for cases where it may possibly infiltrate into the water system. Hydraulic oil is typically filtered and the piping cleaned to a specified level of cleanliness recommended by the equipment manufacturer and furnished with a viscosity of 135 to 165 Saybolt Seconds Universal (SSU), similar to automatic transmission fluid in vehicles. Hydraulic oil systems can be obtained in the low-pressure range of 60 to 300 psig (414 to 2,070 kPa) for smaller valves or in high-pressure systems ranging from 1,000 to 2,500 psig (6,890 to 17,200 kPa) for large valves. Standard heavy-duty hydraulic cylinders have a minimum working pressure rating of 3,000 psig (20,680 kPa). The use of the higher-pressure systems greatly reduces the size of the cylinders and the cost of the hydraulic power system required to operate large valves. The construction of oil power systems is explained in chapter 5.

Hydraulic oil offers superior operating characteristics because it inherently lubricates all of the cylinder and control components that it contacts. Like water, hydraulic oil is not compressible, and therefore it can provide smooth cylinder motion during slow valve operating times.

Hydraulic oil must be ultra clean, especially for high-pressure systems, to avoid clogging or scoring of cylinders and solenoid valve surfaces. Manufacturers may specify a cleanliness level in accordance with industry standards such as ISO 4406. A typical ISO 4406 cleanliness level may be "19/17/14" where the three numbers specify that there shall be less than 5,000 4-micrometer, 1,300 6-micrometer, and 160 14-micrometer or larger particles per 100 milliliter of fluid. The ISO standard contains tables with many levels of cleanliness. The cleanliness level is measured by looking at a small sample of the fluid under a microscope to size and count the particles. New hydraulic oil rarely meets this level; hence, new oil should be filtered before use. Users contemplating a hydraulic oil system should familiarize themselves with other standards and practices concerning cleanliness, such as SAE J1165 and ASTM D4174.

Fluid used for flushing or filling hydraulic systems should be filtered using at least a 10-micron pressure filter system. Hydraulic tanks, fittings, and tubing should also be flushed thoroughly before installation. Once the system is running, it stays clean and trouble-free because of the lubricity and noncorrosive properties of hydraulic oil.

REFERENCES

ASTM D4174, "Standard Practice for Cleaning, Flushing, and Purification of Petroleum Fluid Hydraulic Systems."

ISO 12100, "Hydraulic Fluid Power – Fluids – Method for Coding the Level of Contamination by Solid Particles."

SAE J1165, "Reporting Contamination Levels of Hydraulic Fluids."

Skousen, P.L. 1997. *Valve Handbook*, New York: McGraw-Hill.

This page intentionally blank.

Chapter 2

Principles of Operation

This chapter covers three basic types of hydraulic and pneumatic actuators used in the water and wastewater industry that are included in AWWA C541, Hydraulic and Pneumatic Cylinder- and Vane-Type Actuators for Valves and Slide Gates, latest edition. The mechanics and the physics behind how linear cylinder, rotary cylinder, and vane actuators perform their intended functions are explained so users can determine which actuators best suit their particular needs.

GENERAL DESCRIPTIONS

The primary function of cylinder actuators is to produce linear motion to restrain or operate the closure member of valves and gates. A cylinder actuator consists of a pressure-retaining barrel bolted between head and cap ends, as shown in Figure 2-1. The piston includes seals on its outside diameter to contain pressure between the piston and either end of the cylinder. Important dimensional information for cylinders is the *bore* or inside diameter of the barrel, the *stroke* or maximum travel of the piston, and the *rod diameter*, which changes with cylinder bore and operating pressure.

When supply pressure is directed through one of the ports and applied to one side of the piston, the opposite side vents to drain, creating a differential pressure across the piston. The differential pressure is applied over the area of the piston, resulting in a force or thrust on the rod. It is important to recognize that the areas on either side of the piston are not the same. The piston area on the left side of the figure is reduced by the cross-sectional area of the rod. Hence, a cylinder will produce more thrust when pressure is applied to the cap end than the head end.

In a linear cylinder actuator, the head of the cylinder is rigidly mounted to the valve body and the rod is attached to the valve stem or closure member to mechanically create thrust, operating the valve or gate closure member in a linear manner. Thrust is the linear force on the valve stem or slide gate, generally expressed in units of pound-force (lb-F) or

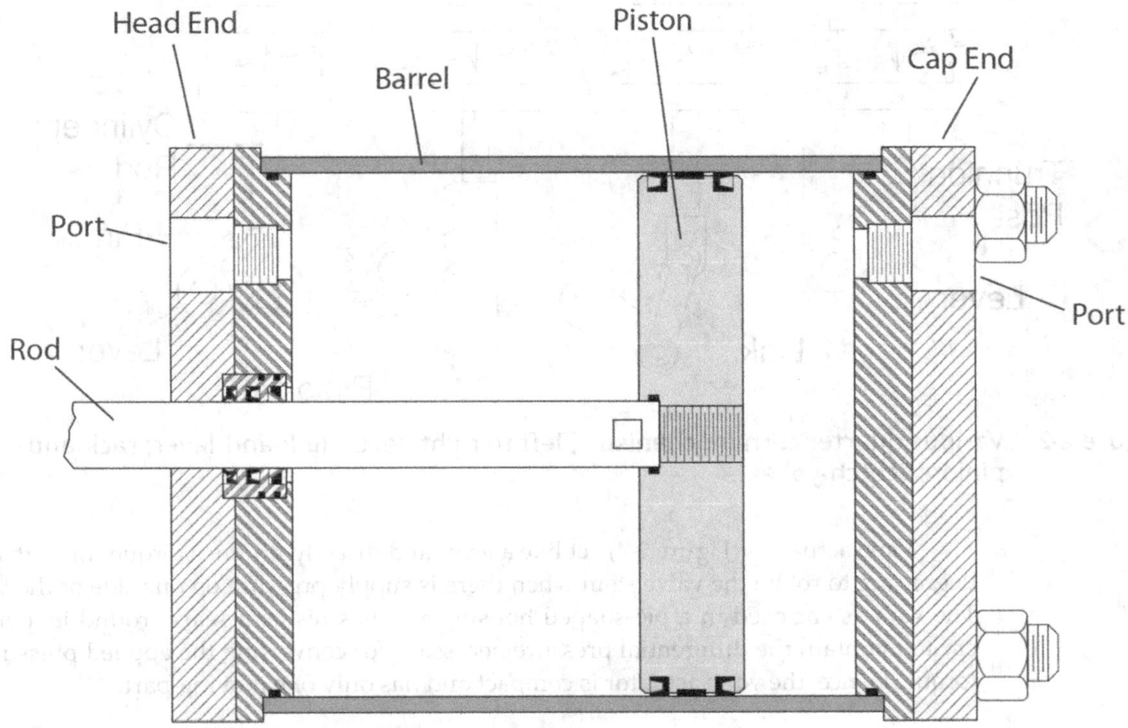

Figure 2-1 Cylinder actuator provides linear thrust

Newton (N). During lifting (opening) of the gate, the head end of the cylinder will be pressurized, with reduced area and reduced thrust.

A quarter-turn cylinder actuator converts the linear cylinder thrust into torque using one of the mechanisms shown in Figure 2-2. Torque is a twisting or turning movement on the valve stem or shaft, generally expressed in foot-pounds (ft-lb) or newton-meters (N–m). The required torque must overcome the frictional and flow-induced forces in the valve to permit mechanical operation of the valve.

The lever mechanism uses a special cylinder whose head pivots about a trunnion or post. The cylinder rod is attached to a simple lever with a rotatable pin. As the cylinder rod extends, it pushes on the pin end of the lever, thereby rotating the lever 90 degrees.

The link-and-lever mechanism is similar except a link is used between the rod and the lever to provide an additional mechanical advantage when the link is perpendicular to the lever. This mechanism produces high lateral side loads on the cylinder rod, so the pin must be supported along its travel by a guide slot in the actuator housing or with a support bar.

The rack-and-pinion mechanism has a straight gear rack attached to the cylinder rod. As the cylinder rod is extended, the rack rotates the pinion gear. This mechanism also produces high lateral side loads on the cylinder rod, so the rack must be supported along its travel.

The scotch-yoke mechanism consists of a cylinder rod and pin that travel in a slot contained within the lever. In most scotch–yoke mechanisms, the cylinder rod is extended across the housing to provide support for lateral side loads.

Another version of the rack-and-pinion mechanism (Figure 2-3) consists of linear racks that mate with a rotating splined-pinion gear inside the cylinder. As air pressure is applied to either end, the pistons in the actuator are sealed against the cylinder housing, driving the two racks and rotating the pinion gear. This type of rack-and-pinion actuator is relatively compact and typically used on smaller valves.

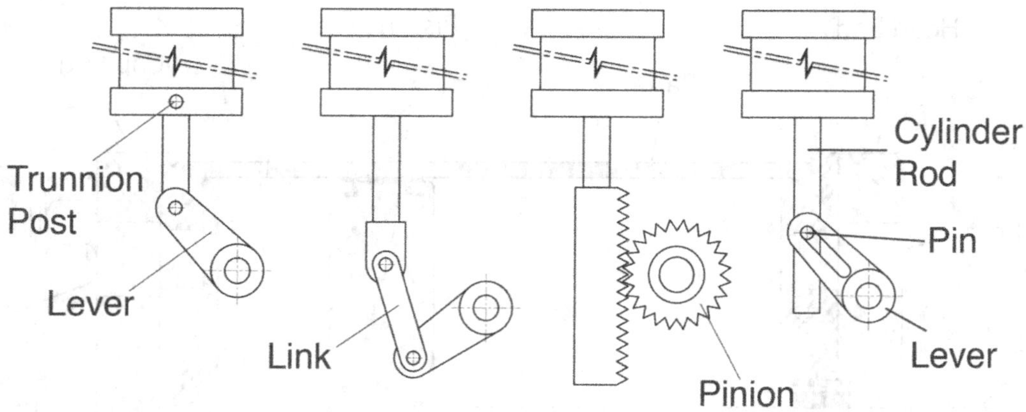

Figure 2-2 Various quarter-turn mechanisms, left to right: lever; link-and-lever; rack-and-pinion; scotch-yoke

Vane actuators (Figure 2-4) act like a lever and directly produce torque through a 90-degree arc to rotate the valve stem when there is supply pressure on one side of the vane. The vane is enclosed in a pie-shaped housing and has resilient seals around its periphery to maintain the differential pressure necessary for converting the applied pressure to torque. Hence, the vane actuator is compact and has only one moving part.

THRUST CALCULATIONS

Slide-gate cylinder actuators are required to generate thrust to lift and lower the slide gate, as shown in Figure 2-5. The thrust should be provided by the gate manufacturer with consideration given to the weight of the gate, the weight of the stem, and the friction on the frame due to the differential pressure acting over the area of gate. Thrust values can vary greatly depending on the types of seats and gate wedges employed in the valve design, so the thrust requirements from the manufacturer are needed to properly size the actuators.

Typically, the maximum thrust required to operate a vertical slide gate occurs while opening the valve, during the release from the wedges and lifting the weight of the gate. The force is in part empirically determined by the manufacturer and normally presented in the following equation (AWWA C560). The 1.5 factor applied to the P_1 term represents the thrust needed to overcome the friction from the wedges used to seat the gate.

$$F = (D_w \times f \times A \times H) + (1.5 \times P_1) + P_2 \qquad \text{Eq 2-1}$$

Where:

F = total force required to open, lb (N)
D_w = weight density of water, 62.4 lb/ft^3 (1,000 kg/m^3)
f = friction factor of slide against the seat = typically 0.35
A = area of the gate opening, ft^2 (m^2)
H = differential head of water at gate centerline, ft (m)
P_1 = weight of slide, lb (kg)
P_2 = weight of stem, lb (kg)

When cylinder actuators are sized for slide gates, a sizing safety factor of at least 1.3 should be applied to the overall thrust.

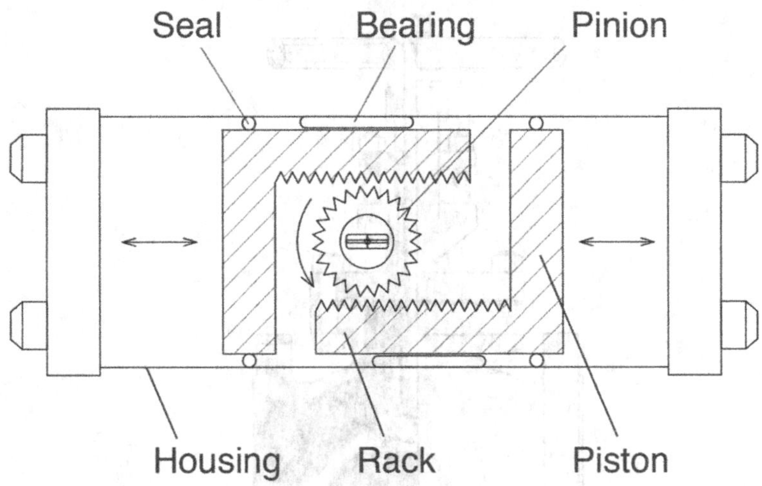

Figure 2-3 Rack-and-pinion mechanism

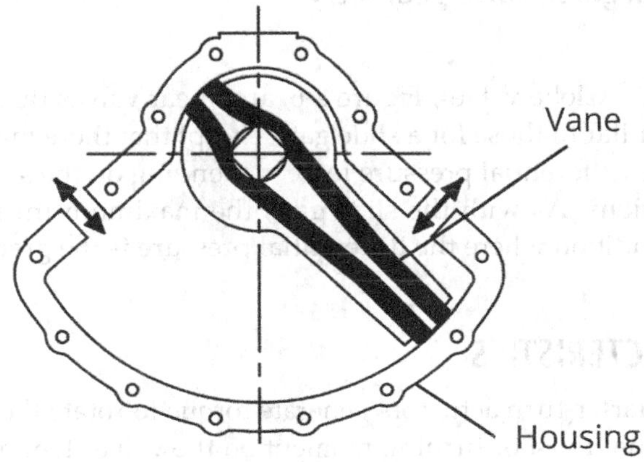

Figure 2-4 Vane mechanism

Figure 2-5 Typical slide gate for waterworks applications

Source: Courtesy of Rodney Hunt

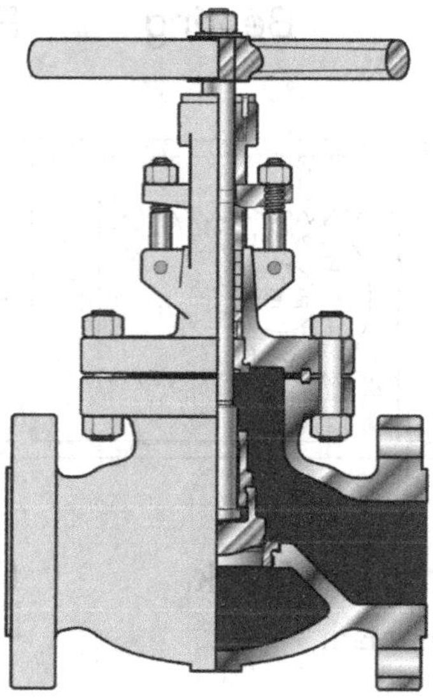

Figure 2-6 Typical globe valve geometry

Globe valves, Figure 2-6, are linear valves that require operating thrust calculations similar to those for a slide gate, except that there may be flow-induced thrust greater than the differential pressure force, depending on the shape of the flow path and the flow conditions. As with the slide gate, the maximum thrust is generated during the unseating condition where the differential pressure is the greatest.

TORQUE CHARACTERISTICS

Quarter-turn actuators generate torque to rotate the valve stem. Torque is defined as rotational twist or turning moment on the valve stem or shaft, expressed in the units of foot-pounds (ft-lb) or Newton-meters (N-m). As can be seen in Figure 2-7, the torque produced by a wrench is found by multiplying the applied force by its distance from the bolt.

Valve torque is generated from several factors within quarter-turn valves that relate to valve design. Some of the torque components and their source include:

- *Seat torque* – friction between the seal and the seat
- *Bearing torque* – shaft friction due to differential pressure on the closure member
- *Packing torque* – seal friction around the valve stem or shaft
- *Dynamic torque* – flow-induced torque occurring in the mid-range of valve operation due to unbalanced differential pressures across the closure member

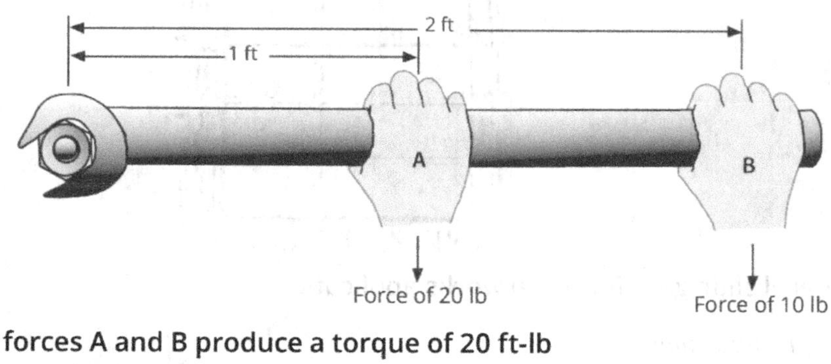

Figure 2-7 Both forces A and B produce a torque of 20 ft-lb

- *Hydrostatic torque* – unbalanced static pressure force across the closure member when only one side of the pipe is full of water and the shaft is horizontal
- *Shaft-offset torque* – unbalanced pressure across an offset closure member
- *Center-of-gravity torque* – unbalanced gravity forces acting on a nonsymmetrical closure member with a horizontal shaft

These torque components are combined to calculate the two required maximum torques for actuator sizing.

Break torque is the amount of rotational resistance that must be overcome to initiate valve motion when the valve is in the closed position. The torque may be higher during seating or unseating of the valve, depending on the valve design. The break torque is typically the sum of the seat, bearing, packing, hydrostatic, shaft-offset, and center-of-gravity torques. In general terms, break torque is proportional to the square of the valve size and the differential pressure.

Run torque is the amount of rotational resistance that must be overcome to sustain motion during valve travel. For larger valves, dynamic torque can play a significant role, and because it varies with valve angle, the maximum run torque is often reported with the valve angle at which it occurs. The run torque is typically the sum of the dynamic, bearing, packing, shaft-offset, and center-of-gravity torques. In general terms, run torque is proportional to the cube of valve size and the square of flow rate. Hence actuators for large valves at high flow conditions will usually be sized on run torque. Run torque is combined with other friction torques to produce the total valve torque, which varies with valve position as shown in Figure 2-8.

AWWA Manual M49, *Butterfly Valves: Torque, Head Loss, and Cavitation Analysis*, provides the methodology for predicting valve operating torque and represents the current method used for the water industry by many valve manufacturers.

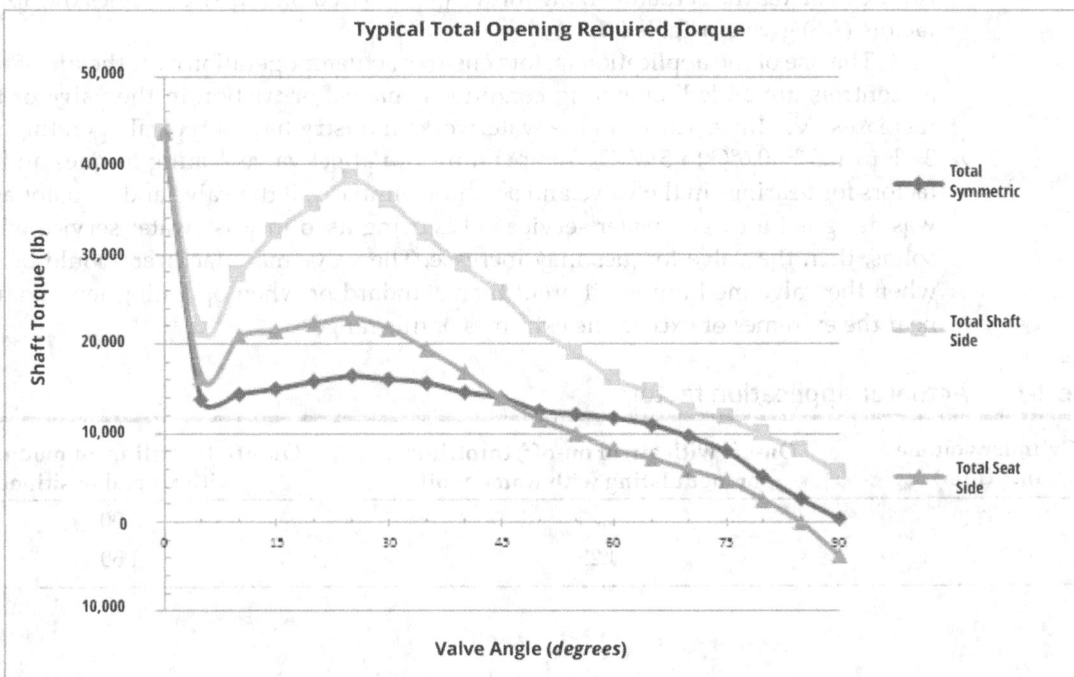

Figure 2-8 Torques for a butterfly valve

POSITION CHARACTERISTICS

It is important to understand the seating characteristics of a valve requiring power actuation. Some types of valves, such as gate valves, are commonly referred to as *torque-seated* because they are seated by applying a specified torque to the stem. Other valves, such as butterfly valves, are commonly referred to as *position-seated* because they are seated by accurately positioning the closure member.

Slide gates are driven into the seat and wedged closed by the actuator. The stroke of the actuator can be adjusted to provide the best seated position to minimize leakage.

Butterfly and ball valves require precise rotational positioning to affect a tight seal, typically within plus or minus one-half of a degree. Certain types of double and triple offset butterfly valves seal tighter as they are closed further into the seat, so the manufacturer will need to specify whether these types of valves should be position seated or torque seated.

Similarly, eccentric plug valves seal tighter as they are closed further. They are sometimes rotated up to 5 degrees past closed to create a seal under high reverse pressures or after plug wear has occurred. The manufacturer's instruction manual should be consulted for the proper method to adjust the closed setting for these valves.

SIZING CONSIDERATIONS

AWWA valve standards require the valve manufacturer to provide the minimum required shaft torque (MRST) values to operate a valve through the entire opening and closing strokes. The MRST values are calculated using the methods described in AWWA Manual M49. The torque at both the seated position (seating and unseating) and the mid-stroke maximum total dynamic run (opening or closing) are evaluated. The actuator manufacturer calculates the actuator sizing torque (AST) based on the MRST times the application factors (AF) given in Table 2-1.

The use of the application factors ensures actuator operation over the life of the valve as controls are added, operating conditions change, or friction in the valve or actuator increases over time. Valves in the waterworks industry have a typical operating range of 33°F to 125°F (0.6°C to 51.7°C). Temperature can affect valve-seating torques and friction factors for bearings in the valve and actuator. Similarly, if the valve and actuator assembly was designed for clean water service but is being used in wastewater service with some solids, then the valve torques may increase. The valve manufacturer should be advised when the valve medium is different than standard or when operating temperatures are near the extremes or exceed the extremes of this range.

Table 2-1 Actuator application factors

Cylinder volume in.^{3} (liter)	On-off with air, or on-off, throttling, or modulating with water or oil	On-off, throttling, or modulating with air and positioner
< 80 (1.3)	1.25	2.00
≥ 80 (1.3)	1.25	1.50

ACTUATOR TORQUE CHARACTERISTICS

Depending on the type of quarter-turn actuator mechanism employed, the torque output versus angle of the actuator position may vary with stroke, as illustrated in Figure 2-9. For the purposes of comparison, all of the torque curves were adjusted to provide the same torque at the 90-degree position. The change in torque output as the actuator is closed can then be comparatively illustrated.

The rack-and-pinion and vane-type mechanisms produce constant torque throughout their travel. Their characteristic curve is shown as a straight horizontal line in the figure.

The scotch-yoke mechanism will typically produce higher torques at both ends of travel. The scotch-yoke curve typically is curved downward in the middle of the stroke, as shown in the figure. Some models may have a canted or slanted slot in the lever, which may provide a greater torque multiplier on one end than the other. If the valve has a higher actuator sizing torque in mid-travel versus at the seat end, then a larger than expected scotch-yoke actuator may be needed.

The link lever will produce higher torques at one end with the higher torque typically used to seat and unseat the valve. Because of higher torque capability on one end than the other, it is important to match the closed position of the actuator with the closed position of the valve, if the increased torque is needed to seat the valve. Link-lever actuators are often economical for valves with high seating torque in the closed position, such as butterfly and ball valves.

The pivoting cylinder actuator will produce higher torques near mid-travel. Pivoting cylinder actuators are often used on large valves where the mid-travel torque is often high because of torques related to flow through the valve.

The actual torque rating of the actuator will depend on its size and supply pressure. The supply pressure varies depending on the supply media system employed. Many applications specify a minimum or range of supply pressures, such as 40 to 80 psig (276 to 552 kPa). In these cases, the actuator sizing should be based on the minimum pressure. Actuator manufacturers publish tables of actuator torque versus supply pressure, as shown in Table 2-2, which are compared to the valve actuator sizing torque.

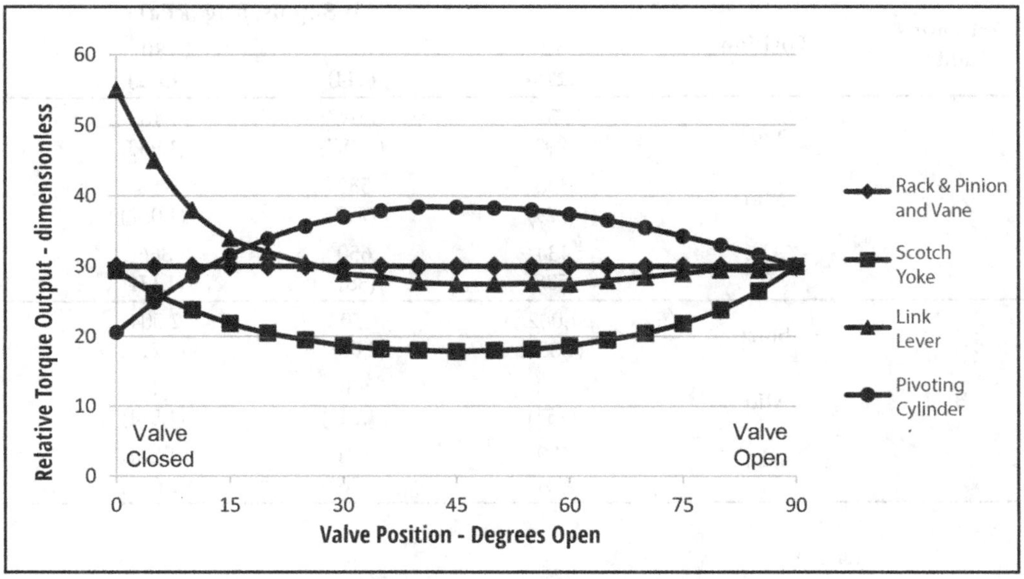

Figure 2-9 Torque characteristics of various cylinder and vane actuators

Pneumatic and hydraulic actuators with springs referred to as *single-acting* require special sizing considerations. The cylinder must create enough thrust to operate the valve while simultaneously compressing the spring. When the spring is needed to operate the valve, be aware that the spring force diminishes as the actuator strokes.

Actuator suppliers publish torque capabilities of their actuators with and without the spring forces (Table 2-2). The torque output is shown for the Start, Mid, and End positions. Start equates to the travel position where the spring is compressed. Mid shows the minimum torque in the center of travel, and End represents the torque at the far end of travel with the spring fully extended. The values shown in Table 2-2 are typical for a scotch-yoke-type actuator where the torque output is less in the Mid positions. Also, the torque is reduced from start to end because as the spring extends and relaxes, its force lessens. It should be noted that as the air supply pressure increases, so does the output torque proportionally. The air supply pressure is limited because using pressures higher than 80 psi (552 kPa) generates torques that exceed the capability of the lever mechanism. This is common for many actuators.

EXAMPLE: Select the single-acting fail-closed actuator model that will operate a butterfly valve with a running torque of 708 ft-lb (960 N-m), a break torque of 1,079 ft-lb (1,463 N-m), and an 80-psig (552 kPa) minimum cylinder supply pressure.

First, a 1.25 safety factor is applied to the torques, yielding a break sizing torque of 1,349 ft-lb (1,829 N-m) and a run sizing torque of 885 ft-lb (1,200 N-m). Second, using the sizing chart in Table 2-2, the run sizing torque 885 ft-lb (1,200 N-m) is compared to the Mid torques listed in the 80-psig (552 kPa) column, and the Model B is shown to generate 1,117 ft-lb (1,514 N-m) at 80 psig (552 kPa) in the Mid positions. Third, the break sizing torque (1,349 ft-lb [1,829 N-m]) is compared to the End torques listed in the 80-psig (552 kPa) column and the Model B is shown to generate 1,241 ft-lb (1,683 N-m) for the end position. The Model B would be selected for this example.

Table 2-2 Typical sizing chart for spring return cylinder actuators

Spring Return Actuator Output Torques, *ft-lb (N-m)*

Actuator Model	Position	Air Supply, *psig (kPa)* 40 (276)	60 (414)	80 (552)	120 (827)
A	Start	732 (992)	1,097 (1,487)	1,463 (1,983)	–
	Mid	390 (529)	585 (793)	780 (1,057)	–
	End	434 (588)	650 (881)	866 (1,174)	–
B	Start	1,002 (1,359)	1,503 (2,038)	2,004 (2,717)	3,006 (4,076)
	Mid	558 (757)	837 (1,135)	1,117 (1,514)	1,675 (2,271)
	End	620 (841)	931 (1,262)	1,241 (1,683)	1,862 (2,524)

Because spring-return cylinders must be sized to overcome the spring force and the valve torque at the minimum supply pressure, the units may become dimensionally large and powerful. There may be a risk to the valve with such a powerful actuator, especially when it is sized at the minimum supply pressure but powered with the maximum supply pressure in the field. Finally, because spring-return actuators can be physically large and heavy, users should verify that there is sufficient space around the final installation to accommodate the actuator and its operation and maintenance. If the valve stem is horizontal, it may be necessary to provide secondary actuator support.

If the valve is obstructed from closing, the actuator may have sufficient torque to exceed the maximum acceptable stem torque (MAST) specified by the valve manufacturer. When critical shutoff valves are specified, the actuator sizing may also need to consider the MAST of the valve.

REFERENCES

AWWA (American Water Works Association). 2009. Standard C541, Hydraulic and Pneumatic Cylinder- and Vane-Type Actuators for Valves and Slide Gates. Denver, CO: AWWA.

AWWA. 2007. Standard C560, Cast Iron Slide Gates. Denver, CO: AWWA.

AWWA. 2012. Manual of Water Supply Practices M49, *Butterfly Valves: Torque, Head Loss, and Cavitation Analysis*, 2nd ed. Denver, CO: AWWA.

Chapter 3

Types of Construction

This chapter presents the construction details of power actuators, including materials selection and the function of components. By understanding the function of the various types of power actuators, the advantages of each type can be understood and the best type properly applied for a given application. This manual does not present all available types of actuators, but covers all of those contained in AWWA C541.

LINEAR CYLINDER

Cylinder-type actuators produce a linear motion to operate slide and knife gates (Figure 3-1). The cylinder consists of two end plates joined together with a barrel and tie rods. Inside of the barrel is the piston, which travels up and down as hydraulic pressure is applied to either end of the cylinder. As differential pressure is applied to the piston, thrust is generated to the piston rod. The cylinder rod is attached to the gate and causes it to rise and fall inside the gate frame (Figure 3-2).

The thrust of the cylinder is computed using the following equations:

$$F_e = \pi D_b^2 P / 4 \qquad \text{Eq 3-1}$$

Where:

F_e = Force during extension, lb (N)
D_b = Diameter of cylinder bore, in. (mm)
P = Supply pressure, psi (kPa)

$$F_r = \pi (D_b^2 - D_r^2) P / 4 \qquad \text{Eq 3-2}$$

Where:

F_r = Force during retraction, lb (N)
D_r = Diameter of rod, in. (mm)

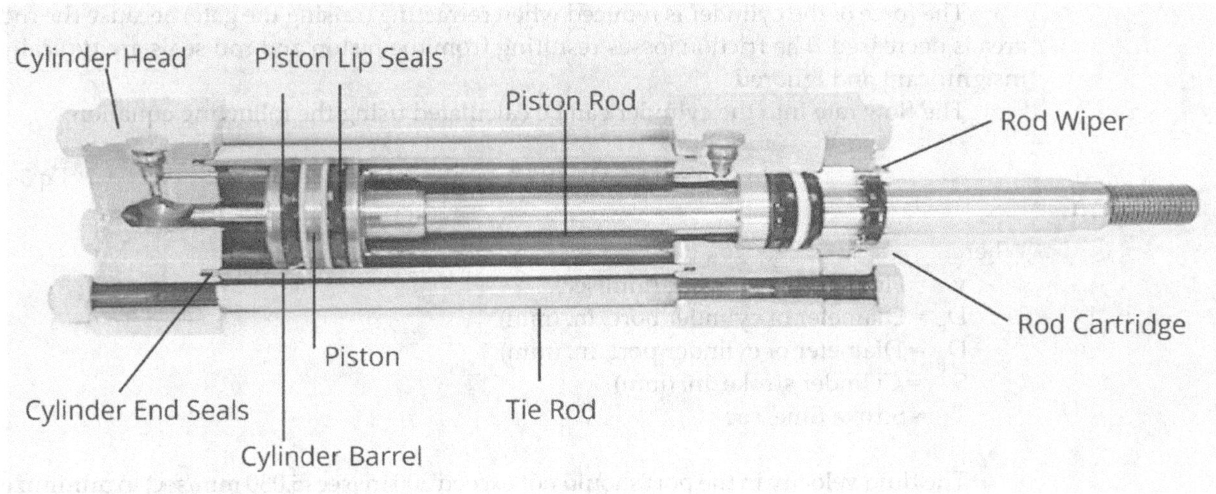

Figure 3-1 Hydraulic cylinder
Source: Courtesy of Milwaukee Cylinder

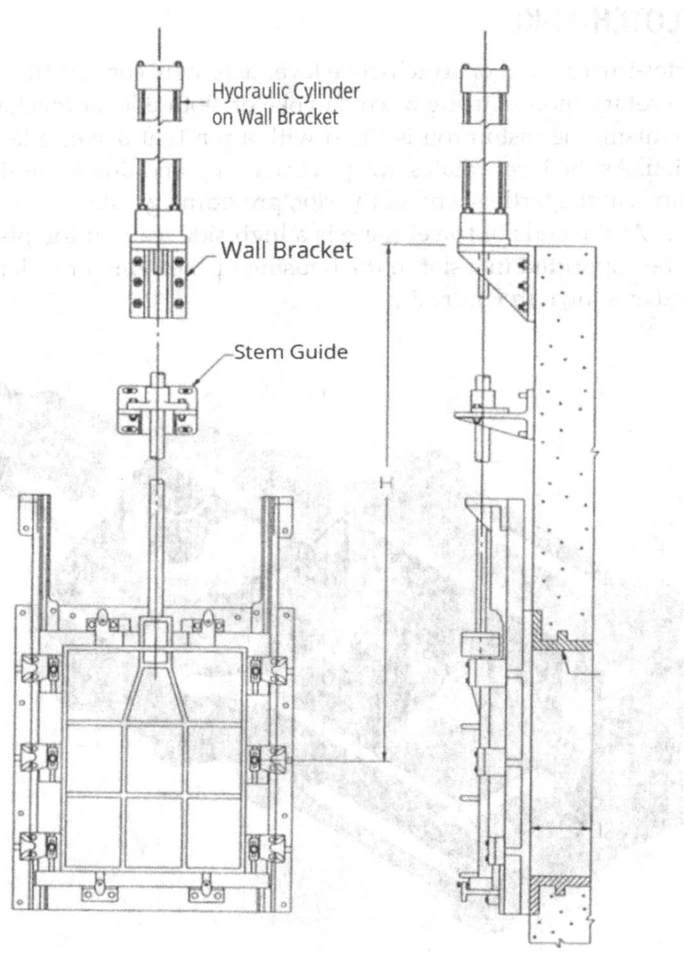

Figure 3-2 Slide gate with cylinder actuator
Source: Courtesy of Rodney Hunt

The force of the cylinder is reduced when retracting (raising the gate) because the rod area is decreased. The friction losses resulting from the piston and rod seals are typically insignificant and ignored.

The flow rate into the cylinder can be calculated using the following equation:

$$v = (D_b / D_p)^2 S / T \qquad \text{Eq 3-3}$$

Where:

v = Fluid velocity, in./sec (mm/sec)
D_b = Diameter of cylinder bore, in. (mm)
D_p = Diameter of cylinder port, in. (mm)
S = Cylinder stroke, in. (mm)
T = Stroke time, sec

The fluid velocity in the port should not exceed 200 in./sec (5,080 mm/sec) to minimize turbulence, head loss, and hydraulic shock. If the desired stroke time (T) results in excessive velocity, then larger ports, larger supply lines, and dual ports should be considered.

QUARTER-TURN SCOTCH-YOKE

Quarter-turn slotted or scotch-yoke lever actuators convert the linear motion of the cylinder to rotary motion using a scotch-yoke or slotted lever mechanism (Figure 3-3). In this mechanism, the piston rod is fitted with a pin that drives a lever through 90 degrees of rotation. As the lever rotates, the pin travels up and down the slot. At both ends of travel, the pin is at the farthest end of the slot, providing greater movement in the arm and more torque. At the ends of travel there is a high side load on the piston rod, so either the pin must be supported in a slot in the housing or the cylinder rod, must be supported on the far end as shown in Figure 3-3.

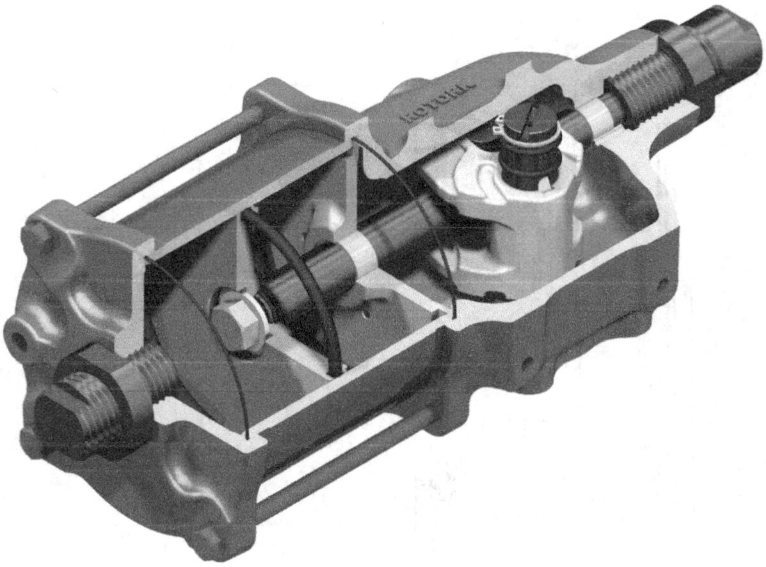

Figure 3-3 Scotch-yoke lever actuator

Source: Courtesy of Flowbus Company

The lever or yoke can have a symmetrical slot or a slanted or canted slot (Figure 3-4). When the slot is canted, the actuator will produce greater torque on one end of travel.

QUARTER-TURN RACK-AND-PINION

Quarter-turn rack-and-pinion actuators convert the linear motion of the cylinder to rotary motion using a rack-and-pinion mechanism in which a linear straight gear is attached to two opposing pistons (Figure 3-5). As the pistons and integral gear racks travel in and out, they drive a round, splined-pinion gear through 90 degrees of travel. The gear racks are on opposite sides of each piston, and the pistons always move in opposite directions from each other. The pistons therefore work together to rotate the pinion gear in the same direction. Rack-and-pinion actuators typically have two pressure ports. One port drives the two pistons apart, and the other drives the pistons together. The housing is precision-bored so the pistons can seal against the inside diameter.

Rack-and-pinion actuators with housing that acts as the cylinder are typically limited to pneumatic service and can be equipped with springs on one or both ends of the housing to allow fail-open or fail-close service when supply pressure is lost. These actuators produce the same torque at all positions unless equipped with a spring. Most units are equipped with stop screws on each end to allow adjustment of the closed position.

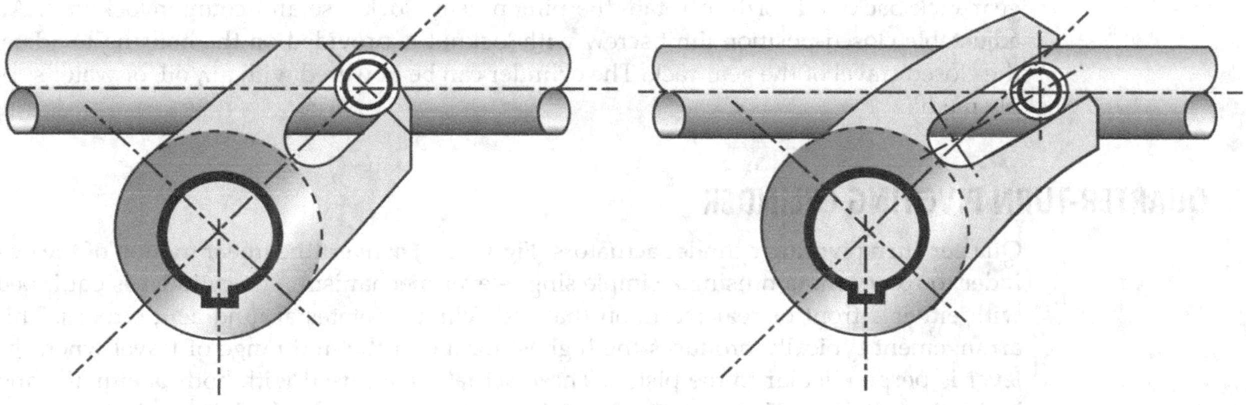

Figure 3-4 Slotted lever mechanisms

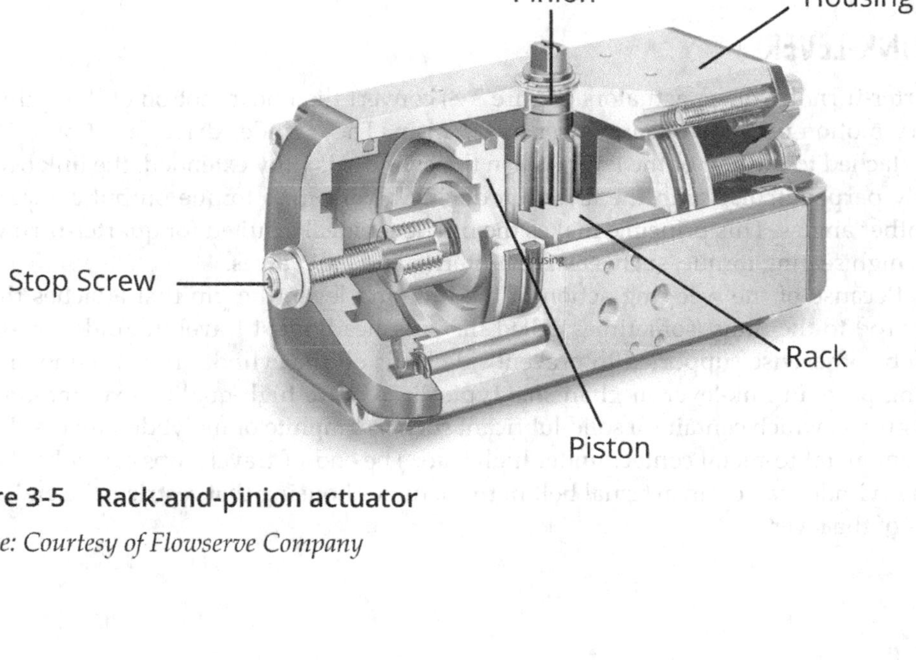

Figure 3-5 Rack-and-pinion actuator

Source: Courtesy of Flowserve Company

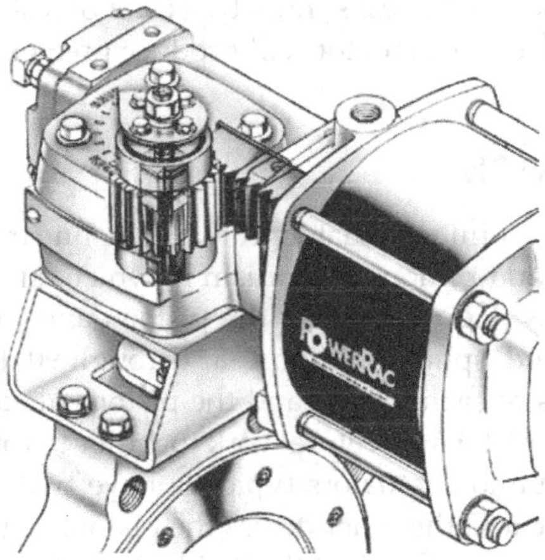

Figure 3-6 Rack-and-pinion actuator with double-acting external cylinder

Source: Courtesy DeZURIK, Inc. Sartell, Minn.-USA

Quarter-turn rack-and-pinion actuators can also be configured with a separate cylinder, as shown in Figure 3-6. In this case, a double-acting cylinder will move a single gear rack back and forth to rotate the pinion gear clockwise and counterclockwise. An adjustable closed-position limit screw with locknut is provided on the housing to adjust the closed travel of the gear rack. The cylinder can be powered with air, oil, or water supply media.

QUARTER-TURN PIVOTING CYLINDER

Quarter-turn pivoting cylinder actuators (Figure 3-7) convert the linear motion of the cylinder to rotary motion using a simple single-lever mechanism. The cylinder is equipped with either a front or rear trunnion that the cylinder rotates around as it strokes. This arrangement typically produces the highest torque in the mid-range of travel when the lever is perpendicular to the piston. These actuators are used with both pneumatic and hydraulic cylinders. The mechanisms of these actuators may not be fully enclosed, exposing external moving parts that may need to be guarded.

QUARTER-TURN LINK-LEVER

Quarter-turn link-lever actuators (Figure 3-8) convert the linear motion of the cylinder to rotary motion using a link-and-lever mechanism. The cylinder drives a set of links that are attached to the end of the lever. When the cylinder is fully extended, the link becomes nearly perpendicular with the lever, producing a very high torque output compared to the other angles. This principle makes the link-lever ideally suited for quarter-turn valves with high seating torque, such as large butterfly and ball valves.

Because of the wedging action of the link and lever, the pin that attaches the cylinder rod to the links (sometimes called the crosshead) must travel in guides or slots or must be otherwise supported to prevent side loads on the cylinder rod. The moving and sliding parts in link-lever mechanisms typically require high-quality, extreme-pressure (EP) grease, which contains a solid lubricant such as graphite or molybdenum disulfide to prevent metal-to-metal contact under high load. The end-of-travel stops can either be part of the cylinder cap or an external bolt in the actuator housing that restricts the end movement of the lever.

QUARTER-TURN VANE

Vane-type actuators produce a 90-degree rotary motion to turn the valve stem and operate the closure member of the valve. However, instead of a cylinder, the vane acts like a lever and generates rotation from the pressure applied to either side of the vane. The vane is sealed against the inside walls of a pie-shaped, split housing with an O-ring or lip-type sealing system. The housing is typically constructed of aluminum or zinc alloy and coated on the interior with Teflon or fusion-bonded epoxy. As pressure is applied to the vane, it pivots about the shaft and produces torque to operate the valve. The torque output is nearly constant at all angles of travel. Adjustable travel stops are typically included in the ends of the housing to prevent overtravel in both directions. These devices are permanently lubricated and should not be powered with lubricated air, which may wash away the lubricant.

While vane actuators are compact and economical (Figure 3-9), they are typically limited to low torque outputs (i.e., 10,000 ft-lb / 13,700 N-m) for small to midsize valves. Vane actuators have externally projecting shafts for transmitting torque, and typically require a shaft coupling and open bracket to mate with the valve.

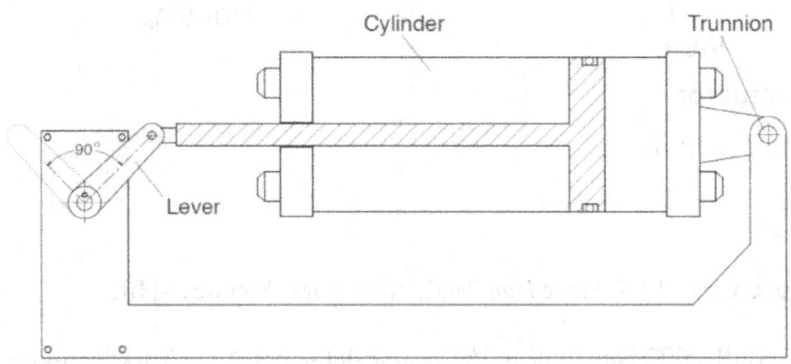

Figure 3-7 Pivoting cylinder actuator

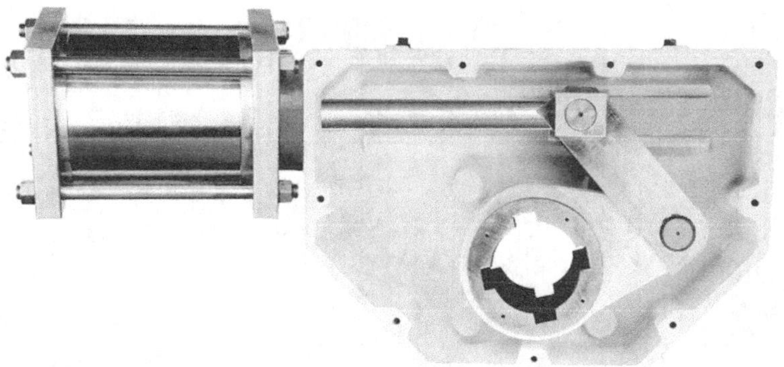

Figure 3-8 Link-lever actuator

Source: Courtesy of Val-Matic Valve & Mfg. Corp.

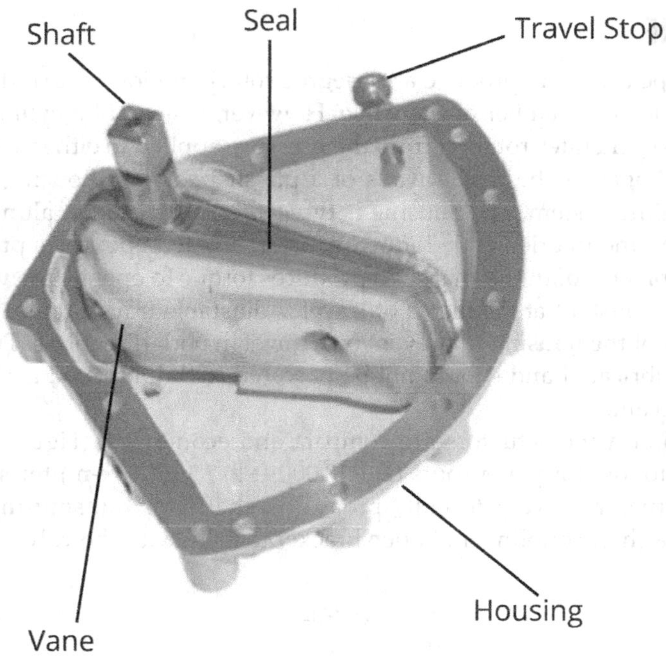

Figure 3-9 Vane actuator

Source: Courtesy of Kinetrol USA

REFERENCES

Skousen, P.L. 1997. *Valve Handbook*, New York: McGraw-Hill.

Nesbitt, B. 2007. *Handbook of Valves and Actuators*, New York: Elsevier.

This page intentionally blank.

Chapter 4

Controls

The action of hydraulic and pneumatic actuators is controlled by directing the supply pressure to either side of the cylinder piston or vane in the actuator. Hydraulic and pneumatic actuators are often specified for unique features they offer over electric motor actuators. For example, hydraulic actuators have speeds of operation that are easily field adjusted, and the open operation can be adjusted independently of the closing operation. Most hydraulic and pneumatic supply systems can be designed to provide excess supply pressure during electrical power outages and can continue to operate the valve to its specified failure position as well.

Numerous types of control elements are available to direct the supply pressure, and some of the more common elements, such as solenoid valves, flow controls, and positioners, will be explained in this chapter.

VALVE CONTROL CONCEPTS

The purpose of a power actuator is to use the air, water, or oil medium to control the position of a valve.

The most basic type of control for valves is on–off service. In this service, solenoids or similar devices direct pressure to the actuator, operating it fully to either open or close the valve and then back again. While on–off service sounds relatively simple, it is important for the purchaser to specify the desired operational speed of the valve because of its potential impact on the performance of the water system or process. For example, while it may be typical to open and close a small valve in 60 seconds, pressure transient conditions may occur if a large valve installed in a long pipeline is closed in a similar short period of time. On long and large-diameter pipelines, valves may be required to close in 5 to 10 minutes. When the flow velocity is stopped within the period of time required for the pressure wave to travel to the end of the system and reflect back (e.g., 15 sec for a 5-mi [8-km] pipeline), it is as if the valve was closed instantly, resulting a significant pressure transient.

A second and more complex type of valve control is position control. Throttling service operation is characterized by the deliberate and infrequent movement of a valve or gate to an intermediate position based on a position control signal to maintain a consistent process. Typically, solenoid valves are used to position power-actuated valves to a specific position for extended periods of time.

If the position signal fluctuates continuously over time, the valve operation is considered modulating service, and more specialized control devices such as positioners or digitally controlled solenoid valves are employed. Positioners can be controlled with either electrical or low-pressure pneumatic position signals. The basic actuator is the same, but the ancillary devices become more complex in this type of operation.

The failure mode of the valve should be specified. There are three basic source or failure indicators:

- Electrical power failure or loss
- Supply pressure failure or loss
- Position signal failure or loss

These indicators must then direct the actuator controls to move the valve to a specified location. The following failure modes are most common:

- Fail open
- Fail close
- Fail in last position

Once the combination of failure indicators and failure modes are determined, the actuator accessories can be selected. Valves can be equipped with supply media storage devices, solenoid valves, positioners, and springs to accomplish many combinations of failure indicators and modes.

CONTROL SYSTEM COMPONENTS

In order to control the operation of the pneumatic or hydraulic actuator, various control elements are needed to direct the supply pressure to and from the cylinder piston or vane in the actuator. Whether the supply medium is air, water, or oil, the functions of control elements such as solenoid valves are basically the same. In pneumatic and hydraulic system design, it is customary to use standard symbols to depict the components in the control system. Figure 4-1 lists some of the most common symbols used with actuators. When the symbols are combined, they form a pneumatic or hydraulic control circuit. The function of some of the components will be discussed, and later in this chapter the operation of some basic control circuits is addressed.

SOLENOID VALVES

A solenoid valve is an electrically powered directional valve that controls the direction of supply medium flow to the actuator cylinder. Solenoid valves are designated as two-way (2 ports), three-way (3 ports), four-way (4 ports), etc. As shown in Figure 4-2, a solenoid valve has a body with multiple ports and one or two electrical solenoid coils that when energized operate the solenoid valve. These valves can also be operated by springs, air, or oil. Typically, a four-way solenoid valve is used to control a double-acting cylinder and a three-way solenoid valve is used to control a spring-return actuator.

A four-way solenoid valve has a supply port, exhaust port, and two cylinder connections (A and B). When the solenoid is energized and deenergized the supply pressure is sent to either the A or B port to operate the cylinder. Hence, when the solenoid is

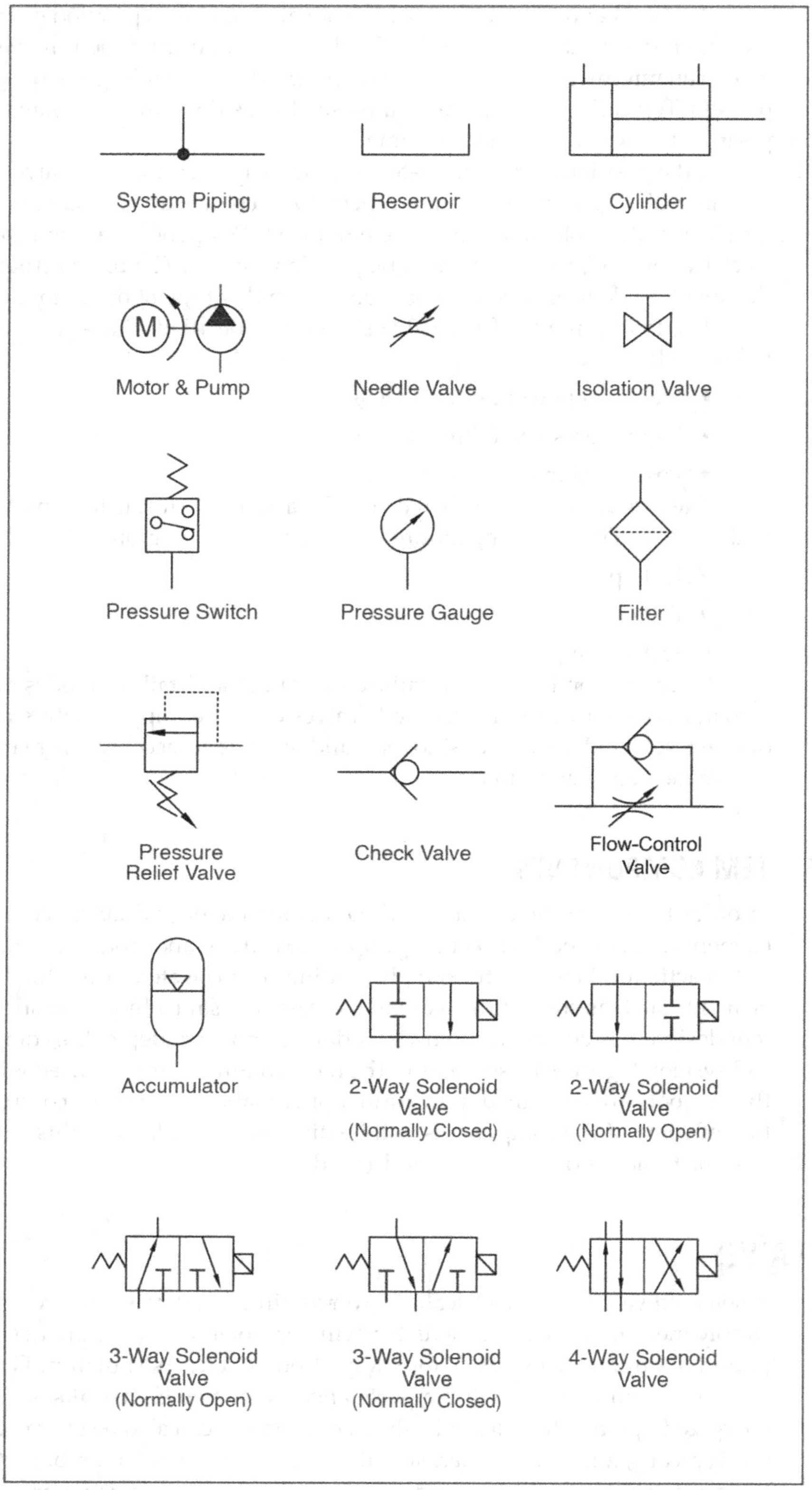

Figure 4-1 Basic hydraulic symbols

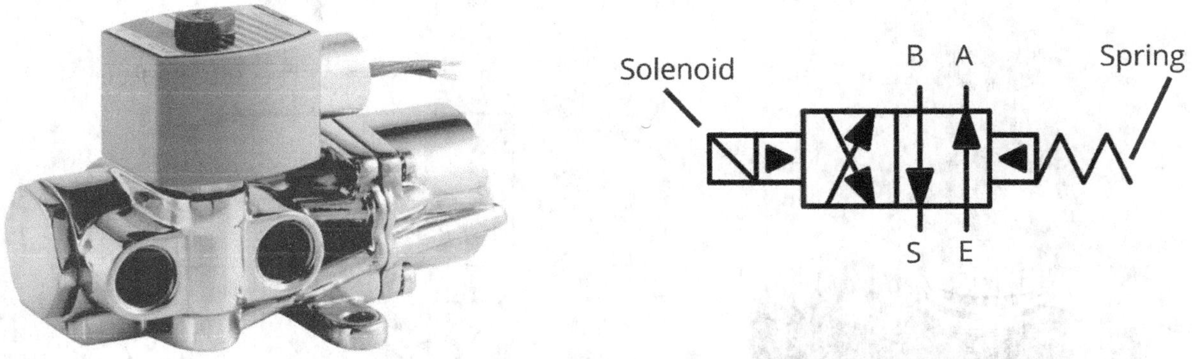

Figure 4-2 Four-way solenoid valve and schematic

energized, the supply pressure will typically open the gate or ball valve. Conversely, when the solenoid is deenergized (or loses electrical power), the supply pressure will close the valve or gate.

The deenergized position is referred to as the *fail-safe* position. System designers can use solenoid valves and specify the fail-safe position (open or closed) of various valves in the plant when electrical power is lost. If the valves were equipped with electric motor actuators, they would fail in their last position. Special solenoid valves are available with a blocked center position that can also hold a valve in the last position after a power outage.

The solenoid can be energized by various electrical supply voltages; 24-volt DC and 115-volt, single-phase AC are the most common. Because the electrical solenoid represents an electrical hazard, the solenoid enclosures are furnished in many optional configurations to provide a degree of protection to personnel against incidental contact and some protection against specified environmental conditions. The National Electrical Manufacturers Association (NEMA) and International Electrotechnical Commission (IEC) publish standards such as NEMA Publication 250 for electrical enclosures; common ratings listed include the following:

- NEMA Type 1 (IP10) – General indoor use
- NEMA Type 3 (IP54) – General outdoor use
- NEMA Type 4 (IP56) – Watertight
- NEMA Type 4X (IP56) – Watertight and corrosion resistant
- NEMA Type 6 (IP67) – Submersible
- NEMA Type 6P (IP67) – Submersible for prolonged periods
- NEMA Type 12 (IP52) – General indoor use with gasketed door

There is no direct relationship between NEMA and IP (Ingress Protection) ratings. The exact requirements should be obtained from the equipment supplier, NEMA, or IEC.

FLOW-CONTROL VALVES

Cylinder actuators should be equipped with flow-control valves as shown in Figure 4-3. A flow-control valve is used on double-acting cylinders and has two parts: a needle valve portion and a check valve portion. In one direction (controlled flow), the flow rate out of the cylinder is controlled by a needle in the flow-control valve. In the opposite direction, a check valve allows free flow past the needle valve into the cylinder. Flow-control valves are typically installed on both ports of the cylinder. The valves are oriented so that the

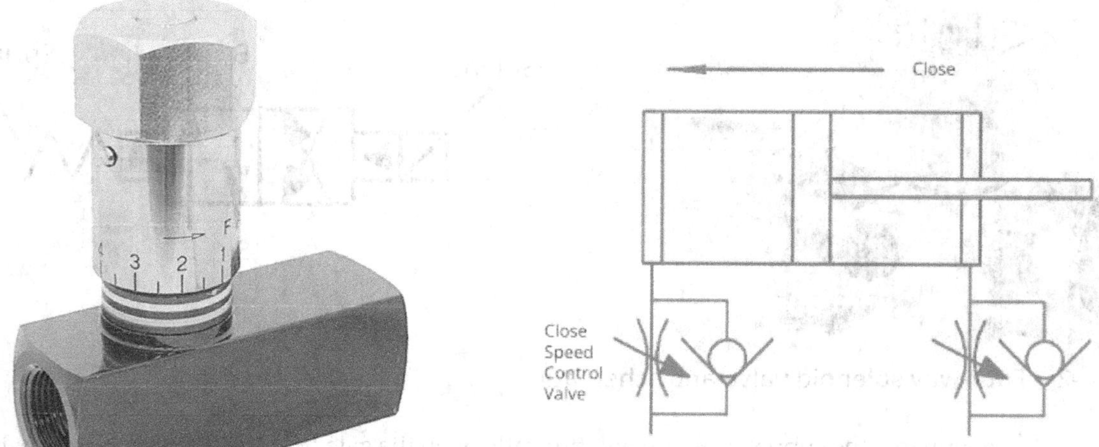

Figure 4-3 Flow-control valve

flow out of the cylinder is controlled; hence, the "Close" speed-control valve is shown on the left in the hydraulic schematic in Figure 4-3. That way, the valves control the fluid in the high-pressure side of the cylinder and, concurrently, control the movement of the piston. It should be noted if the flow into the cylinder is controlled, the gate or ball valve may drive the actuator forward and cause a vacuum to occur in the cylinder, thereby preventing smooth operation. Consequently, flow-control valves should be installed with the "Controlled Flow" arrow pointing away from the cylinder.

CONTROL CIRCUITS

The control devices previously discussed are piped together to control an actuator in thousands of different configurations. Two basic control circuits detailed in the following section demonstrate how to read schematic diagrams.

Figure 4-4 illustrates a control circuit for operating a valve equipped with a double-acting hydraulic cylinder actuator. The cylinder may be connected directly to a valve gate or may be connected to an intermediate gear box mechanism. The pump/motor—(3) in the figure—draws hydraulic oil from a reservoir (1) through a filter (2) and charges an accumulator (6) until the pressure switch (4) turns off the pump motor. After the pump is stopped, the pressurized fluid will be maintained in the accumulator because of the check valve (5). If the pressure switch fails, any excess oil will be returned to the reservoir through the relief valve (7). The pressurized oil is directed through the deenergized four-way solenoid valve (8) to the rod end of the cylinder (9) and closes the valve. When the solenoid valve is energized, it shifts to the opposite position, directing the pressurized oil to the cap end of the cylinder and opening the valve. Upon loss of electrical power, the solenoid valve shifts back to the position shown, and stored oil in the accumulator will perform according to the programmed failure mode.

Figure 4-5 illustrates a similar control circuit except for the addition of flow-control valves (12a and 12b), a needle valve (10), and a two–way solenoid valve (11). When solenoid valves (8) and (11) are energized, oil is directed from the accumulator (6) through the check valve side of the flow-control valve (12b) to the cap end of the cylinder (9) and open the valve. The speed of opening will be controlled by the needle portion of the flow-control valve (12b). When the four-way solenoid valve (11) is deenergized as shown, flow is directed through the check valve portion of flow-control valve (12a) to the rod end of the

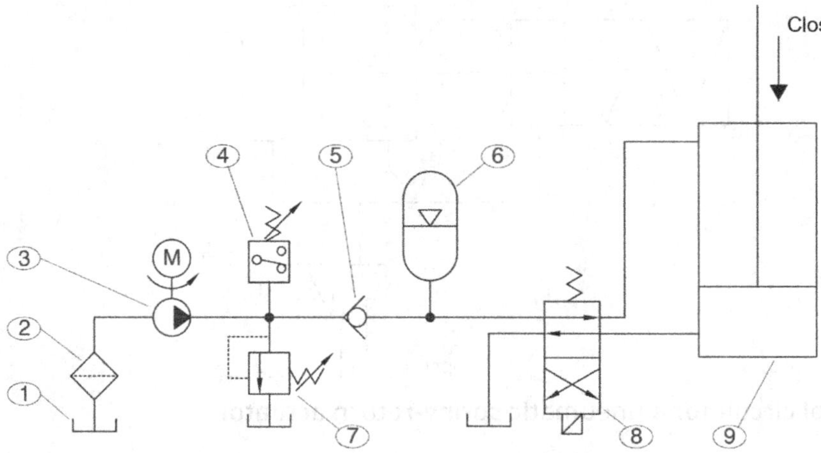

Figure 4-4 Basic control circuit for double-acting cylinder

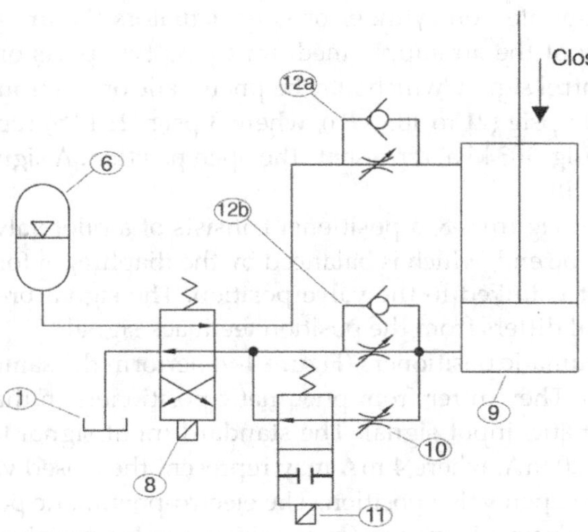

Figure 4-5 Basic control circuit for double-acting cylinder with timed stroke

cylinder closing the valve. The speed of closure will be controlled by the needle portion of flow-control valve (12b). Upon loss of electrical power, both solenoid valves will be deenergized (as shown), directing oil from the accumulator to the rod end of the cylinder and causing the actuator to enter the programmed failure mode. However, with both solenoid valves deenergized, the speed of closure is controlled by both flow-control valve (12b) and needle valve (10). This circuit provides independent speed control of routine opening and closing, and closing after electrical power outage.

Figure 4-6 shows a typical control circuit for a pneumatic single-acting spring-return actuator. Single-acting cylinders are powered from only one end, and the operating force in the return direction is a spring. A three-way solenoid valve (1) is used for single-acting cylinders. When deenergized, the three-way solenoid valve releases the air from the cylinder, allowing the spring to operate the valve. The speed of closure may be controlled by a needle valve (2).

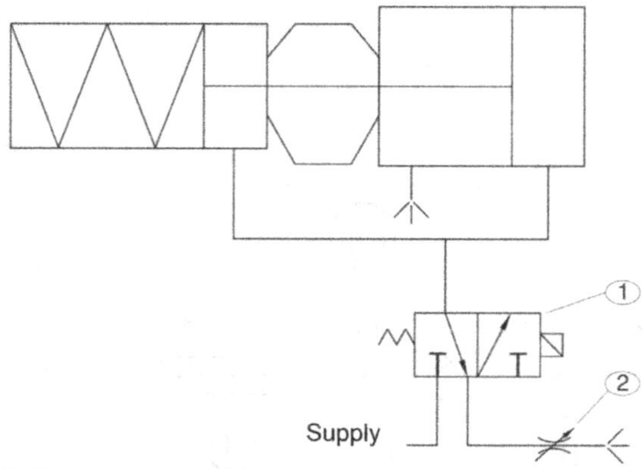

Figure 4-6 Control circuit for a pneumatic spring-return actuator

POSITIONERS

Positioners are mounted on cylinder or vane actuators (Figure 4-7) and are designed to regulate the flow of the air supply medium to the two ports of the cylinder or vane in response to a control signal, which may be pneumatic or electronic. The air control signal is typically 3 to 15 psig (21 to 103 kPa), where 3 psig (21 kPa) represents the closed valve position and 15 psig (103 kPa) represents the open position. A signal is used to position the valve proportionally.

As shown in Figure 4-8, a positioner consists of a pilot valve that is controlled by a range spring on one end, which is balanced by the diaphragm force on the other end. The feedback element is linked to the valve position. The signal pressure operates the pilot valve as the signal differs from the position feedback signal.

Electro-pneumatic positioners (Figure 4-9) perform the same basic function as pneumatic positioners. They differ from pneumatic positioners in that they use an electrical, instead of pneumatic, input signal. The standard input signal for an electro-pneumatic positioner is 4 to 20 mA, where 4 mA may represent the closed valve position and 20 mA may represent the open valve position. The electro-pneumatic positioner employs analog circuitry and an electronic means for measuring valve position (potentiometer or electronic sensor). An electrical output from the positioner's circuitry is sent to an onboard transducer that drives a pneumatic valve, which in turn controls the actuator/valve proportional to the input signal.

Smart positioners (Figure 4-10) typically operate on the same input signal (4 to 20 mA) as electro-pneumatic positioners. Smart positioners maintain valve position control through a microprocessor embedded in the electronics. A smart positioner typically has an enhanced user interface (display and keypad). Operator error is minimized and calibration is made simpler and consistent with the use of auto-calibration routines that are handled by algorithms in the microprocessor. Algorithms in the processor also result in improved performance of the control valve package with respect to speed of response, accuracy, and repeatability. Some smart positioners have shifted away from the analog input signal and are controlled with digital bus networks such as Fieldbus and Profibus. These networks feature a single cable or bus to communicate to many devices digitally through a network, eliminating the need for individual control wires to each valve.

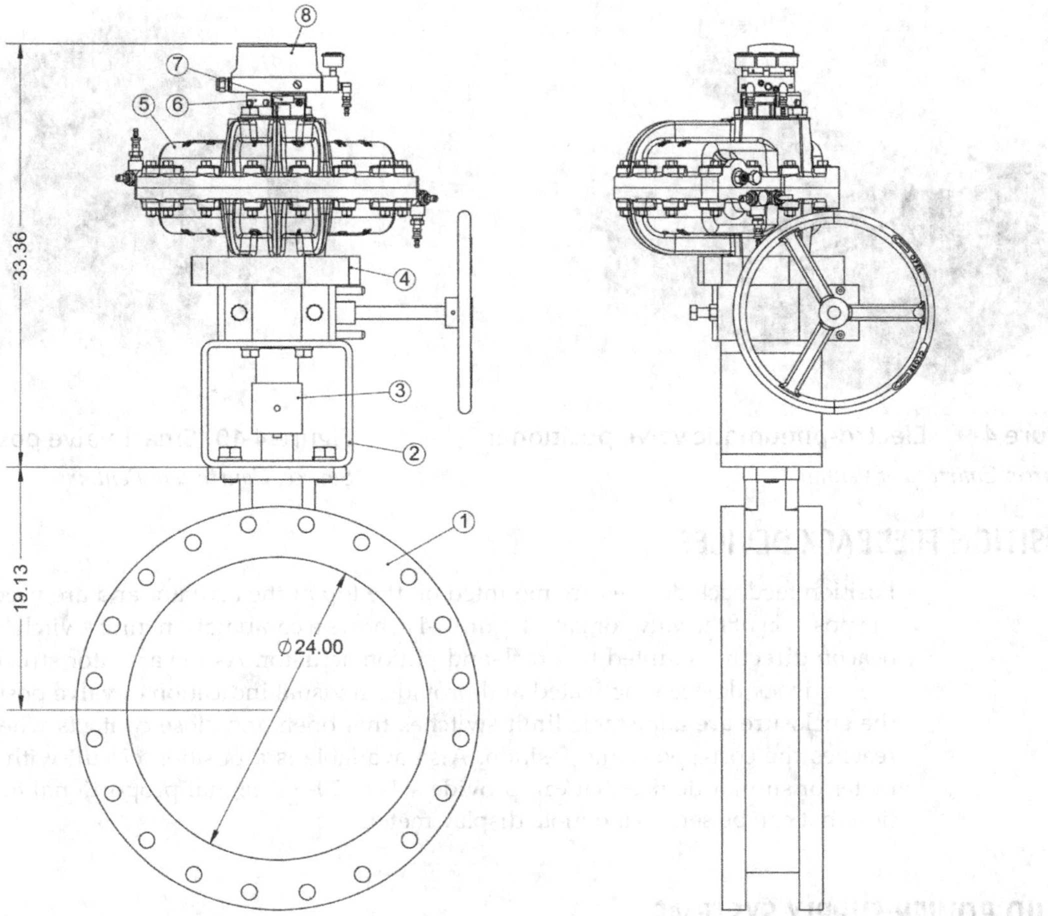

Figure 4-7 Valve positioner, vane actuator, and handwheel override worm gear unit mounted on a butterfly valve

Source: Courtesy of Rotork

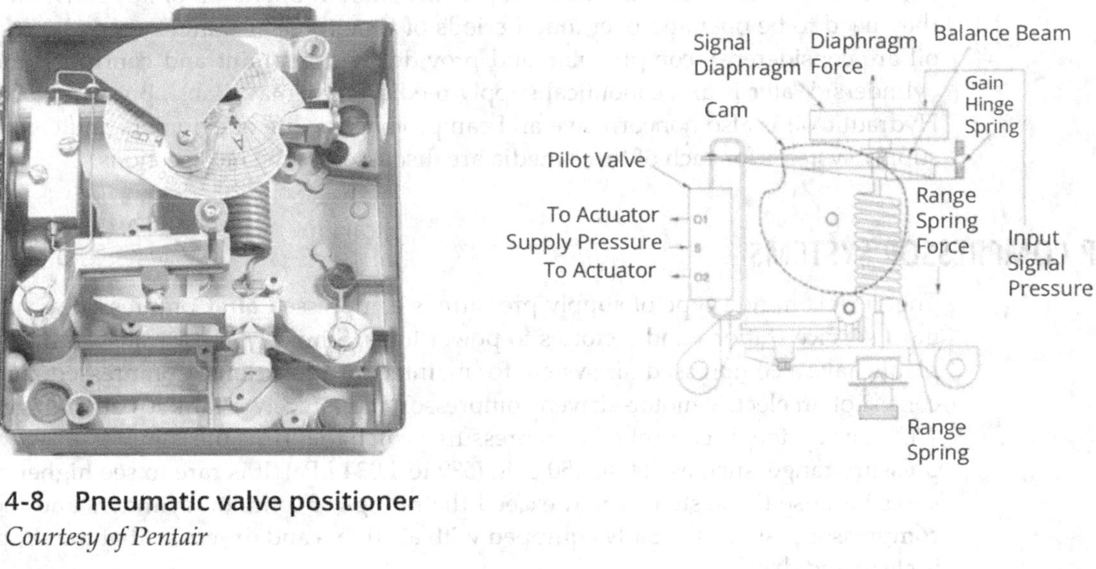

Figure 4-8 Pneumatic valve positioner

Source: Courtesy of Pentair

Figure 4-9 Electro-pneumatic valve positioner
Source: Courtesy of Pentair

Figure 4-10 Smart valve positioner
Source: Courtesy of Pentair

POSITION FEEDBACK DEVICES

Position feedback devices are mounted on the top of the actuator and are used to display the position of the valve or gate. Figure 4-11 shows a combination limit switch housing and beacon directly mounted to a rack-and-pinion actuator. As the actuator strokes or turns, the feedback device is activated and provides a visual indication of valve position. Inside the enclosure are adjustable limit switches that open and close contacts when the valve reaches the corresponding position. Also available is a position circuit with a potentiometer or similar device that can provide a 4- to 20-mA signal proportional to valve position that can be sent to a remote display meter.

FLUID POWER SUPPLY SYSTEMS

Cylinders and vanes can be operated with various types of supply medium including air, water, and hydraulic oil pressure. Depending on the application, each medium has its advantages. Air is probably the most common supply medium because it is commonly used for maintenance in plants, and mechanics are very familiar with its use and properties. Air is compressible and may not provide smooth operation of large cylinders when they need to be operated over long periods of time (greater than 60 seconds). Water and oil are considered incompressible and provide rigid restraint and controlled motion of cylinders. Water is an economical supply medium when available, but can be corrosive. Hydraulic oil is also noncorrosive and can prolong the life of cylinders and controls. The supply systems for each of three media are discussed in the next sections.

AIR COMPRESSOR SYSTEMS

The most common type of supply pressure is compressed air. Compressed air is used at most service stations and factories to power tools. Similarly, most water and wastewater plants have a compressed-air system for maintenance functions. Compressed-air systems consist of an electric motor–driven compressor and a receiver tank for air storage (Figure 4-12). The system is controlled by a pressure switch that runs the compressor within a set pressure range, such as 100 to 150 psig (689 to 1,034 kPa). It is rare to see higher air pressures because the system would exceed the ratings of common cylinders and tools. The compressor system is usually equipped with air filters and dryers so that the supplied air is clean and dry.

Figure 4-11 Valve feedback unit with position beacon

Source: Courtesy of Valworx, Inc.

Figure 4-12 Air compressor system

Source: Courtesy of Gardner Denver

A typical air distribution system is illustrated in Figure 4-13. The system is fed by the duplex air compressor supply unit shown in Figure 4-12. The pressurized air is stored in the air receiver and held within a set pressure range to provide a smooth on and off sequence for the compressor motors. Following the receiver, dryer and filter units are typically used to clean and dry the supply air. A regulator unit may be used to precisely control the pressure of the air supply so that valves and equipment perform in a consistent and repeatable manner. An auto drain is a device that opens automatically to release trapped condensate in the system. The system then will have one or more supply headers connected to the valve actuators around the plant. Each actuator has an isolation control to facilitate maintenance.

HYDRAULIC OIL SUPPLY SYSTEMS

A typical hydraulic oil supply system is shown in Figure 4-14. Hydraulic oil is often specified because it provides reliable long periods and variable speeds of operation. It reduces maintenance of controls and cylinders in the plant because of its noncorrosive properties.

Hydraulic oil systems consist of electric motor–driven oil pumps and air compressors mounted on a reservoir or return oil tank. The oil pumps deliver pressurized oil to an air-over-oil accumulator tank until a level switch has been satisfied. Then, air compressors controlled by pressure switches come on and charge the air space over the oil to the system operating pressures, up to 300 psig (2,070 kPa) but typically 150 psig (1,034 kPa). The accumulator tank will then supply a steady supply of oil, even after an electrical power outage.

Hydraulic oil systems can also be supplied for high-pressure service at 1,500 psig (10,340 kPa) or even 2,500 psig (17,200 kPa). In high-pressure systems, nitrogen-charged accumulators must be used (Figure 5-8) instead of air-over-oil tank to prevent the mixture of high-pressure oil and air, which can be explosive. High-pressure oil systems are practical for large valves because the size of the hydraulic actuator cylinder can be reduced, and subsequently so can the volume of the oil in the hydraulic system. In both low and high oil-pressure systems, the returned oil flows through filters to maintain system cleanliness.

The hydraulic oil system includes an electrical control panel, which contains the motor controls and system logic. On the door of the cabinet are displays of the pump and compressor operations, the system pressure, and alarm conditions. HAND, OFF, and-AUTO switches are typically provided to locally control the operation of the pumps and compressors. The HAND position is used to turn on and test each motor individually.

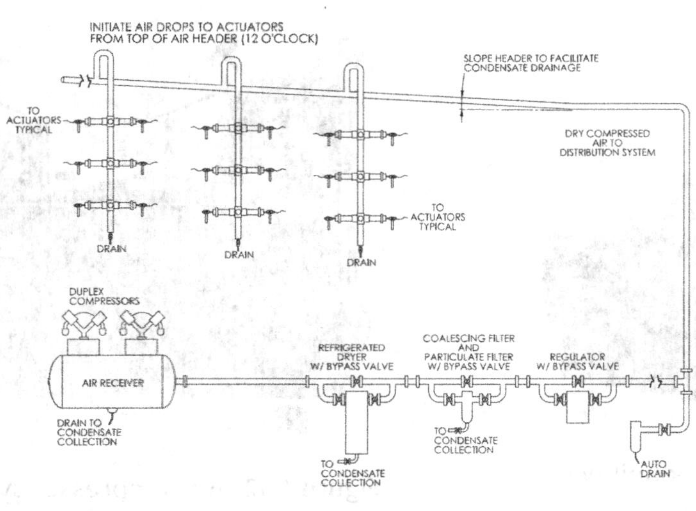

Figure 4-13 Air distribution system

Source: Courtesy of Rotork

Figure 4-14 Low- and high-pressure oil hydraulic system

Source: Courtesy of Val-Matic Valve & Mfg., Corp.

The OFF position allows the user to isolate a motor for service while the system remains operational. The AUTO position is the normal position, and it enables the compressors and pumps to operate automatically, maintaining the oil level and air pressure in the storage tank.

ELECTRO-HYDRAULIC ACTUATORS

Electro-hydraulic actuators are linear or quarter-turn actuators equipped with an integral hydraulic power system. Different designs are available. The electro-hydraulic actuator depicted in Figure 4-15 operates on a pump-and-bleed principle wherein a motorized oil pump provides hydraulic pressure in one direction and the spring-return in the opposite (bleed) direction. The flow of oil to the actuator is controlled by electric solenoid valves and can provide both open–close or modulating operation in response to a control signal.

Another electro-hydraulic actuator design features a totally enclosed system consisting of a power unit with integral hydraulic reservoir, actuating unit, nitrogen gas–charged piston accumulator, positive displacement pump with electric motor, solenoid valves, and limit switches. These components are mounted in a sealed housing unit.

With any of these designs, the actuator can cause the valve to fail in the open, close, or last (current) position, on loss of electrical power.

WATER SUPPLY SYSTEMS

A water supply system for actuator operation is prevalent in water treatment plants and water pumping stations. The water plant or pumping station usually has pumps that deliver pressurized water at 70 to 100 psig (483 to 689 kPa) to the distribution system. It is a simple matter to tap the water supply and provide a filtered pressurized water source for the hydraulic actuators. A clean water supply may not be available at some remote wastewater pumping stations or in wastewater pumping stations that do not need a clean water supply for items such as pump seals. In this case, a small booster pump and accumulator tank may be used, as shown in Figure 4-16. Two considerations when using water are that water may be corrosive, requiring more expensive actuators and controls, and the water pressure may not be available during plant startup, which would require a temporary water supply.

Figure 4-15 Electro-hydraulic quarter-turn actuator

Source: Courtesy of Rotork

Figure 4-16 Water supply system

REFERENCES

IEC Publication 60529, Classification of Degrees of Protection Provided by Enclosures.

NEMA. 2003. Publication 250-2003, Enclosures for Electrical Equipment.

This page intentionally blank.

Chapter 5

Special Equipment

Special equipment can be provided to operate the actuator without supply pressure, including handwheels, hand pumps, and springs.

HANDWHEEL OVERRIDE

Hydraulic actuators can be supplied with a separate manual override to allow actuation without electrical or fluid power. One manual method is to use a worm gear with a declutch mechanism (Figure 5-1). When the horizontal bar is raised, it engages the worm against the segment gear and allows the handwheel to drive the valve open or closed. To employ this method, procedures to isolate the supply pressure to the cylinder and bleed-off pressure on both ends of the cylinder must also be in place. Otherwise the handwheel will be working against the torque of the cylinder actuator.

Pneumatic or hydraulic cylinders can be equipped with a jackscrew consisting of a handwheel and threaded rod affixed to the end of the cylinder (Figure 5-2). When there is no supply pressure, the jackscrew will push the cylinder's piston and operate the valve. Some double-acting cylinder actuators can have a jackscrew on each end to allow manual operation in both directions.

Pneumatic and hydraulic cylinders can also be equipped with an additional hydraulic cylinder, hand pump, and directional valve affixed to the main cylinder (Figure 5-3). When there is no supply pressure, the manual pump pressure is used to stroke the main cylinder. The existing travel stops in the actuator will continue to function as they do with normal operation.

LOCKOUT MECHANISM

The Occupational Safety and Health Administration (OSHA) and other similar regulatory safety authorities publish standards to address the practices and procedures necessary

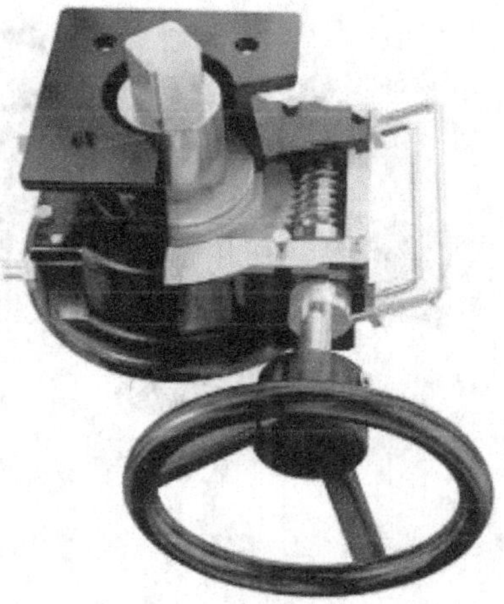

Figure 5-1 Worm-gear manual override with declutch mechanism

Source: Courtesy of Cameron

Figure 5-2 Jackscrew manual override for cylinder

Source: Courtesy of Pentair

to disable machinery or equipment, thereby preventing the release of hazardous energy while employees perform servicing and maintenance activities. The standards outline measures for controlling and isolating hazardous energies—electrical, mechanical, hydraulic, pneumatic, chemical, thermal, and other energy sources.

When service is required on a valve and a segment of a pipeline must be isolated, a lockout mechanism is needed. Valve safety lockouts are an essential workplace mechanism to protect workers from injuries resulting from the accidental opening or closing of actuated valves. There are many types of lockouts available, but in general, the device should lock the actuator in position, and the device should have the ability to be clasped and padlocked to restrict equipment access and facilitate safety (Figure 5-4).

SPRING RETURN

Hydraulic and pneumatic actuators can be equipped with springs to provide fail-open or fail-close operation with loss of electric power and supply pressure. For example, in Figure 5-5, when the cylinder piston strokes to the left the supply pressure operates the valve and compresses the spring. In many cases a spring cylinder is much longer than the actual stroke of the actuator. Long springs provide a more uniform force as they are compressed. Upon loss of electrical power, a solenoid valve releases the air in the cylinder and the spring operates the piston to the right, turning the valve shaft a quarter turn. When the spring-return feature is specified, it is important to specify whether the valve needs to fail open or fail closed so that the actuator can be mounted in the correct orientation on the valve.

Vane actuators can be equipped with a rotary spring unit, similar to a clock spring mounted in a separate sealed housing as shown in Figure 5-6. These springs are designed to open or close the valve automatically without pneumatic power. They have low torque loss through the 90 degrees of operation and allow the valve to fail open or closed as stipulated. When a spring is specified, the sizing of the unit must consider the need to wind the spring and operate the valve at the same time.

Figure 5-3 Hand pump manual override for cylinder

Source: Courtesy of Pentair

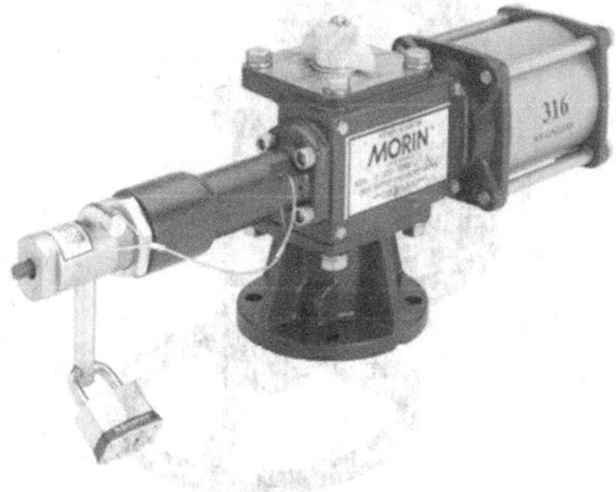

Figure 5-4 Example of lockout mechanism on cylinder actuator

Source: Courtesy of Pentair

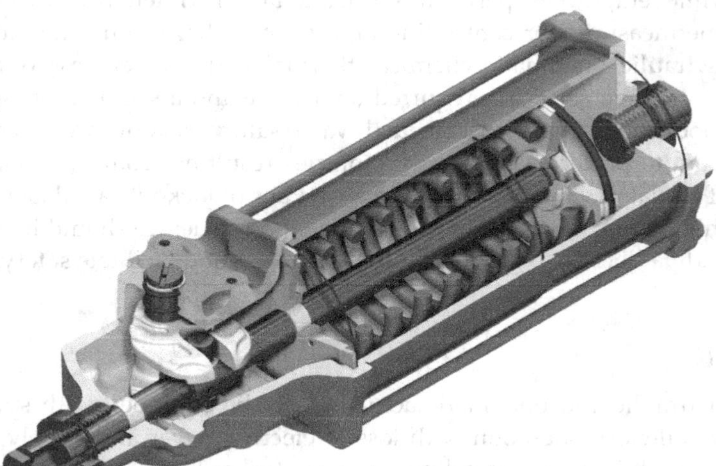

Figure 5-5 Cylinder actuator with spring return

Source: Courtesy of Flowbus Company

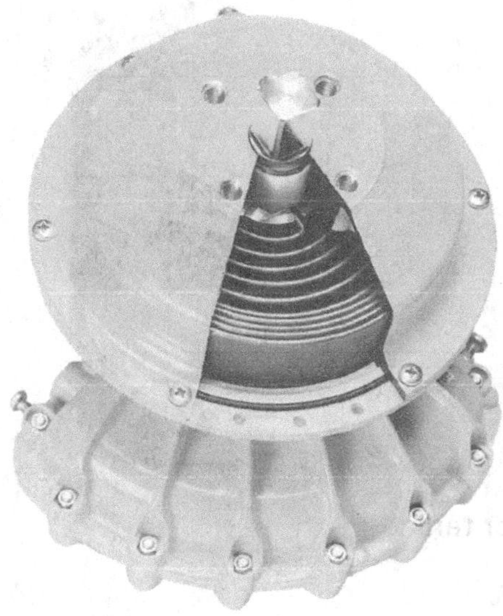

Figure 5-6 Vane actuator with spring return

Source: Courtesy of Kinetrol USA

ACCUMULATORS

When electrical power is lost, an accumulator can supply pressurized media to the actuator, allowing fail-safe operation of the valve. For water supply applications, a well tank with an elastomer bladder precharged to the required minimum pressure, i.e., 30 psig (207 kPa), can be used (Figure 5-7). A similar tank without a bladder can be used for air supply service and provided as an accessory or integrated into the actuator. After a power outage, the water tank or air receiver will supply water or air to operate the actuator for at least one stroke to the fail-open or fail-close position as required.

Pneumatic actuators can be supplied with air fail-safe tanks to provide a ready supply of compressed air to position the valve in the event of an air supply failure. The fail-safe tanks can be sized to provide one or multiple valve operations, can be set to fail the valve in the close or open position, and are often field reversible.

For hydraulic oil actuators, high-pressure accumulators store media under pressure, as shown in Figure 5-8. An accumulator consists of a body or vessel and an internal bag or piston to isolate a gas from the media at the bottom. As pressurized fluid is pumped into the bottom of the accumulator, the gas is compressed and acts like a spring to return the pressurized fluid to the system when needed. An inert gas such as nitrogen is typically used as the energy storage medium to prevent the potential for combustion with the hydraulic oil. The gas is typically injected at 100 psi (689 kPa) less than the minimum operating pressure of the hydraulic system to prevent bottoming out of the bag or piston. Some actuators can be equipped with an integral accumulator for fail-safe operation upon loss of electrical or hydraulic power.

There are numerous formulas, charts, and computer programs readily available for sizing accumulators. To size an accumulator, the initial gas pressure, minimum operating pressure, maximum operating pressure, and required oil volume are entered into ideal gas law formulas to produce the required volume of the accumulator.

Figure 5-7 Water accumulator tank

Source: Courtesy of Flexcon Industries

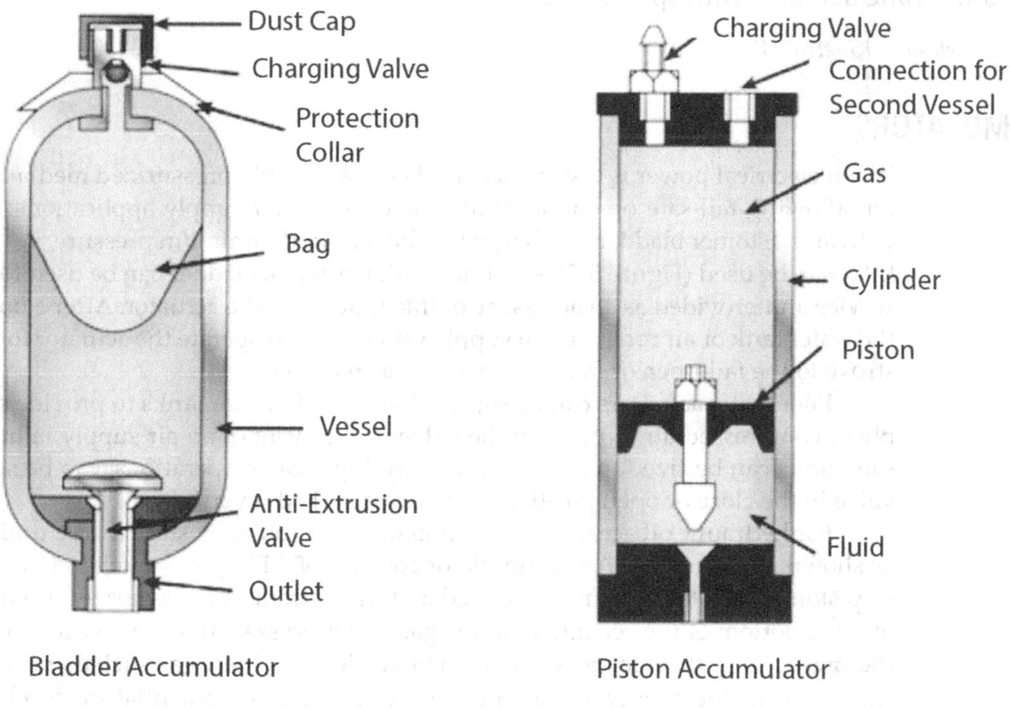

Figure 5-8 Hydraulic oil accumulators

Source: Courtesy of Hydraulics & Pneumatics *magazine, © Penton Media Inc.*

This page intentionally blank.

Chapter 6

Installation and Adjustment

To allow the valve actuator to properly operate, reasonable care is needed in handling, installation, and maintenance. This chapter provides the basic instructions for working with cylinder and vane actuators, but it is important that the instructions provided with the actuator be carefully reviewed and followed. Often, the actuator requirements are specified in accordance with the actuator data form from AWWA C541 Hydraulic and Pneumatic Cylinder and Vane-Type Actuators for Valves and Slide Gates, as shown in part in Figure 6-1, which can be used to help install or troubleshoot an actuator.

STORAGE, HANDLING, AND MOUNTING

Inspect valves and actuators for damage upon receipt prior to offloading from shipment. Unload valves and actuators carefully to the ground without dropping. Lift valve with straps or hooks in the flange holes. Do not lift valves with slings or chains around the actuator or through the valve seat area. Extra care must be taken when handling cylinder actuators with controls and control piping that can be easily damaged.

Valves and actuators should remain crated, clean, and dry until installed to prevent weather-related damage. The valve may be shipped slightly open to maintain the resilient seat in the unloaded condition. For long-term storage greater than six months, indoor storage is recommended.

The instruction manual supplied by the manufacturer should be reviewed in detail before the actuator is installed. At the job site prior to installation, the actuator should be visually inspected and any packing or foreign material in the ports of the actuator should be removed. The nameplate information on the actuator should be verified to ensure the actuator is the one specified.

Customer Information	Valve Information
	Size/Type: ____
	Rotary ____ Linear ____
	Maximum Required Torque or Thrust / units: ____
	Rotation of Rotary Valve
	Clockwise to close ____
	Clockwise to open ____
	Degrees from close to open ____
Project Name / Location	Travel of Linear Valve (close to open): ____ inches
	Valve Stem/Shaft Type & Diameter: ____
	Weight of Gate & Stem for Slide Gate: ____
	Quantity: ____ Tag No(s). ____

Actuator Information

Rotary (check one)		OR	Linear (check if applicable)	
Piston Cylinder	____		Piston Cylinder	____
Rack and Pinion	____			
Scotch Yoke	____			
Vane				

Driver Medium (check one): Pneumatic ____ Hydraulic (water) ____ Hydraulic (oil) ____

Minimum / Maximum Driver Medium Pressure Available at Point of Use (specify): ____ **PSIG**

Failure Mode on Loss of Pneumatic or Hydraulic Pressure (check one)
Fail Free ____ Fail Close ____ Fail Open ____ Fail in Last Position ____

Figure 6-1 Actuator data form in part from AWWA C541, Appendix A

Using the required clean supply medium and pressure, the actuator should be stroked fully before installation on the valve to ensure proper operation. During the test stroke it is important to make note of the direction of rotation and the orientation of the mounting bolts and shaft key. Also, if the actuator is equipped with a spring, its position with the spring relaxed must be noted so that it will match with the designated fail position of the valve, which may be fail-open or fail-close. This information will allow proper fitting to the valve; for example, if the actuator is to be installed with the valve in the closed position, then the actuator should be moved to the closed position to match the valve's position.

Mounting of the actuator is critical to its proper operation. First, the direction of rotation must be compatible with the valve operation. Valves such as plug valves may be eccentric and must be closed by turning the valve stem clockwise. Hence, if the actuator includes a spring for fail-close operation, it must be configured accordingly. Second, the actuator should be mounted to clean dry surfaces free of oil, grease, and preferably paint. The mounting bolts must firmly clamp the actuator to the valve, developing sufficient friction to transmit the actuator torque. See the manufacturer's instruction for proper mounting and bolt torques. If the actuator is mounted loosely, then it may rotate during operation and the closed stop adjustments will not be effective in controlling the valve seated position.

One methodology for confirming the rigidity of the actuator mounting is provided in Manufacturer's Standardization Society standard MSS SP-101. The torque capability of the mounting bolt pattern can be computed by assuming a working stress of the mounting bolts to be 30,000 psi (206.8 MPa) and a coefficient of friction between the mounting surfaces (dry iron or steel) to be 0.3. The following equation can be used:

$$T = f\, S_b\, A_b\, N_b\, D_{bc} \div 2 \qquad \text{Eq 6-1}$$

Where:

T = frictional torque, lb-in (N-m)
f = coefficient of friction, given as 0.3
S_b = bolt stress, given as 30,000 psi (206.8 MPa)
A_b = bolt tensile stress area, in.^{2} (cm^2)
N_b = number of bolts
D_{bc} = diameter of bolt circle, in. (cm)

If the actuator mounting rotates on the valve during operation, then Equation 6-1 should be used to compare the mounting bolt torque capability to the required operating torque of the valve. If the bolt pattern is insufficient, it may be necessary to supplement it with dowel pins or keys tightly inserted through the mounting interface.

The actuator may be attached to the valve shaft with a coupling or extended bonnet. After mounting, the actuator should be cycled three times at a minimum to ensure proper operation with no binding of the coupling or extended bonnet. If the actuator is equipped with a spring, verify that without pressure, the valve fails to the correct position. The pressure required to operate the valve and compress the spring should be noted and compared to the minimum required supply pressure.

SETTING OF THE TRAVEL STOPS

The valve open and closed positions can typically be adjusted by turning the limit screws in the side of the actuator housing and securing the setting in place with a lock nut. Other units may require adjusting a bolt in the center of the cylinder; there may be a bolt on each end of the cylinder to provide both open position and closed position adjustment. Ball and butterfly valves typically require the actuator to precisely position the valve closure member in the center of the seat to within one half of a degree. Other types of valves such as eccentric plug valves may require over travel (up to five degrees) to shut the valve tightly for specific applications. The valve manual should be consulted for these types of valves. Figure 6-2 illustrates the two external screws with locknuts used to set the open and closed positions for a quarter-turn cylinder actuator.

Figure 6-2 Open and closed adjusting screws on a ball valve hydraulic actuator

REFERENCES

AWWA. 2008. AWWA Standard C541-08 Hydraulic and Pneumatic Cylinder and Vane-Type Actuators for Valves and Slide Gates. Denver, CO: AWWA.

Manufacturer's Standardization Society (MSS). 2012. SP-101-2012, Part-Turn Valve Actuator Attachment—Flange and Driving Component Dimensions and Performance Characteristics. Vienna, VA: MSS.

Skousen, P.L. 1997. *Valve Handbook*. New York: McGraw-Hill.

Chapter 7

Maintenance and Safety

To ensure safe and long operation of an actuator and supply pressure system, reasonable care is required to maintain the equipment. This chapter provides safety instructions and maintenance recommendations for hydraulic and pneumatic actuators. When using hydraulic and pneumatic actuators, it is important that the instructions provided with the actuator be carefully reviewed and followed.

REGULAR MAINTENANCE

The most important maintenance item for pneumatic and hydraulic actuators is keeping the supply medium clean. Inspect the operating medium on an annual basis for dirt, scale, discoloration, and other signs of deterioration. Change medium filters on a regular schedule. Verify that the system is providing the supply medium at the proper pressure to the actuator. The required pressure should be marked on the actuator or noted in the owner's file on the equipment.

Most actuators require a check of the lubricant on a regular (at least annual) basis. Check the supplier's instructions on the correct grade of grease to be used. Checking and replacing the grease may require the removal of the actuator cover. Some actuators (particularly vane-type) may be permanently lubricated; owners should consult the instruction manual.

For proper operation of the valve, maintaining a firm and rigid mounting connection of the actuator to the valve is crucial. After many valve operations, actuators tend to work themselves loose. To determine this, observe the operation of the actuator and check the alignment to see if the housing is rotating during operation. If it is, then the actuator mounting bolts must be tightened.

To maintain the proper sealing ability of the valve, the closed position setting of the actuator should be checked to ensure the valve is sealing as required. If the valve begins to leak, the actuator mounting and closed stop settings should be checked.

If the actuator is equipped with flow-control valves, the operating time of the valve should be checked annually to verify that it is within the owner's requirements. If the valve appears to be operating too slowly, the flow-control valves may be partially clogged or the cylinder may need maintenance. The flow-control valves should be opened fully during a portion of the stroke, taking care to avoid inducing transients in the piping system to flush out debris, and then reset.

Refer to other maintenance items noted by the manufacturer, such as regular lubrication and specific settings.

Caution must be used when working on a cylinder since it is a pressure-containing vessel. Removing parts while under pressure could cause personal injury or equipment damage. Release the pressure from both ends of the cylinder before attempting to disassemble or repair. Once apart, the seals and sealing surfaces can be inspected for damage or wear. The manufacturer will provide specifications for lubrication and reassembly, including bolt torques.

TROUBLESHOOTING

Some problems and solutions are presented here to assist in troubleshooting the valve and actuator assembly in an efficient manner. Consult the supplier's manual for specific troubleshooting guidance.

WARNING: Most cylinder actuators are pressure vessels and present serious hazards, so all warnings posted on the valve and actuator should be understood before proceeding.

- *Leakage at cylinder ends:* Tighten tie rod nuts or replace head seals (Figure 7-1).
- *Noise during operation:* A creaking noise during air operation may indicate a lack of lubrication (dry air). Refer to the product instructions about adding lubrication to the air supply or spray some oil into the cylinder ports. Some actuators, such

Figure 7-1 Cylinder head with tie rod nuts
Source: Courtesy of Milwaukee Cylinder

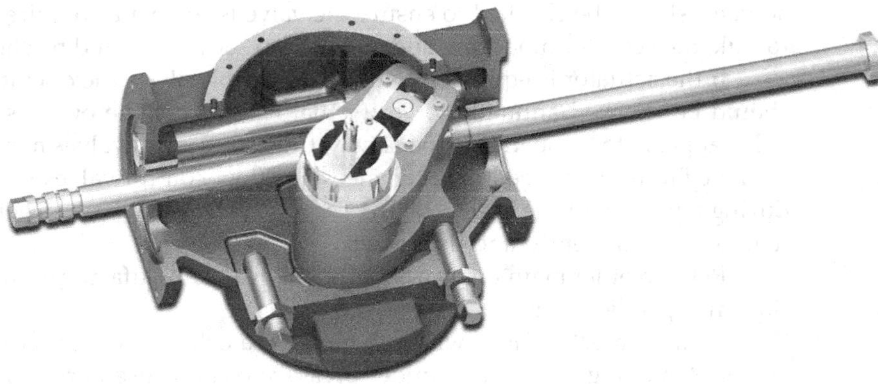

Figure 7-2 Cylinder actuator limit screws

Source: Courtesy of Bray International, Inc.

as vane actuators, are permanently lubricated at the factory, and the addition of lubricants may be detrimental to the operation of the actuator.

- *Valve leaks when closed:* Adjust the stop limit screws (Figure 7-2) by first opening the valve slightly to remove the load from the stop limit screw. Adjust screws one-quarter turn at a time.
- *Valve does not open:* Verify adequate cylinder supply pressure by reading the power supply system pressure gauge. The actuator nameplate or valve documentation should indicate the minimum operating pressure required to operate the valve.
- *Leaking oil at fittings:* Tighten cylinder fittings. Add thread sealant or thread tape if necessary.
- *Valve does not respond to position signal:* Check the supply pressure, the positioner or solenoid valves, and the seals in the cylinder. A lack of actuator response may also be the result of increased torque required by the valve. Verify the valve conditions. The valve components and flow conditions often change over time.
- *Valve does not fail to the required position:* Verify that the air pressure is being exhausted by the controls. Safely check the actuator spring by following the instructions provided with the unit. WARNING: It is important to release the spring energy when actuators are taken apart to avoid bodily harm.

SAFETY CONCERNS

Pneumatic and hydraulic actuators are pressure vessels with a maximum pressure rating indicated on the actuator. As such, the air, oil, or water pressure supply system must never exceed the rated working pressure of the actuator. The pressure supply system must contain fail-safe mechanisms (e.g., relief valves) that limit the maximum pressure supplied to within safe levels.

As stated earlier, when actuators are equipped with springs it is important to carefully release the spring energy by slowly unbolting the cap that holds them down while actuators are taken apart to avoid bodily harm.

The valve and drain line should be opened before removing or repairing the actuator or the valve and the actuator may suddenly move, causing fluid loss or bodily injury. Drain the piping system and place the valve in the closed position to remove the actuator.

The power supply to the cylinder should be shut off and the connection removed, and the cylinder should be drained before working on the cylinder. Otherwise, bodily injury or damage to property may result. Similarly, when a spring-return cylinder is disassembled, the spring may be storing significant energy that when released can cause bodily harm. Always review the manufacturer's instruction manual to disassemble a cylinder actuator.

Working on valves in enclosed spaces requires a knowledge of the rules of confined spaces. Consult Occupational Safety and Health Administration (OSHA) or other regulatory safety authorities' rules and procedures, such as the need for harnesses and ground-level supervision, for all underground work.

Glossary

accumulator A cylindrical vessel in which fluid is stored under pressure to provide a source of fluid power

actuator A device or mechanism used to restrain or move a slide gate or valve closure member

application factor (AF) A safety factor used as a multiple of the valve torque for sizing actuators

break torque The turning force required to fully seat or unseat a valve

bus network A digital communication system with a single main cable or bus that is connected to various industrial control devices through a chained, ring, or branched wiring arrangement

clock spring A flat spring wound in the shape of a spiral used to produce torque

cylinder actuator A fluid-powered actuator that uses hydraulic or pneumatic pressure within a cylinder or housing to restrain or move a valve closure member

cylinder bore Inside diameter of a cylinder or actuator housing

double-acting A cylinder actuator that requires fluid power in both the open and closed directions

eccentric plug A plug whose journal centerline is offset from the centerline of the valve seat

electro-hydraulic actuator A linear or quarter-turn hydraulic actuator equipped with an integral hydraulic power system

failure mode (fail open, close, or in last position) The automatic response of the actuator upon loss of electrical power, supply pressure, or position signal

flow-control valve A small instrument valve that combines the functions of a needle valve and a check valve to precisely control supply medium flow in one direction and allow free flow in the opposite direction

fluid power The use of a pressurized liquid or gas to drive an actuator

fluid velocity The speed at which a fluid travels through a tube or pipe

globe valve A valve that has a round opening to let liquid pass and that closes when a stem presses a disk against the round opening

handwheel A rimmed component designed to facilitate manual actuation of a valve

hydraulic cylinder A device that converts fluid pressure acting on a piston within a cylinder into a mechanical force

hydraulic oil supply A power source consisting of oil supplied and delivered under pressure

linear cylinder actuator An actuator that uses a hydraulic cylinder to generate thrust and operate a valve or gate in a linear fashion

link-and-lever actuator A quarter-turn mechanism in which a link is used between the cylinder rod and the lever to provide an additional mechanical advantage when the link is perpendicular to the lever

lockout A safety device used to lock an actuator and prevent operation of the valve or slide gate while the piping system is under service

maximum acceptable stem torque (MAST) The twisting force that would cause permanent damage to the valve

modulating Operation characterized by the continuous positioning of a valve or slide gate between fully open and closed, in response to a continuous control signal

pivoting cylinder actuator A quarter-turn mechanism in which the rod of a trunnion-mounted cylinder is attached to a lever

position feedback The transmission of the position of a valve visually or through an electrical or pneumatic position signal

position signal An electrical (typically 4–20 milliamp) current or pneumatic (typically 3–15 psi) pressure used to remotely indicate valve position

potentiometer An electrical device with variable resistance based on rotation used to generate a 4–20 milliamp position signal

rack-and-pinion actuator A quarter-turn mechanism in which a piston drives a straight gear rack in a linear manner to rotate a pinion gear

run torque The full load or dynamic torque required to operate a valve or slide gate between the fully open and fully closed positions

safety factor The ratio between an actual property of a component and that required under design conditions to minimize risk and to provide for unknown or unforeseen events that can adversely affect service life

scotch-yoke actuator A quarter-turn mechanism in which a cylinder rod uses a pin that travels back and forth in a slot contained within a rotatable lever

solenoid valve A small directional device or valve used to control the flow of a supply medium to and from a cylinder or vane actuator and operated by an electrical current that energizes a cylindrical coil of wire acting as a magnet

stroke Movement of a valve or slide gate closure member from the fully open to the fully closed position or reverse

supply medium A pressurized fluid used to drive cylinder and vane actuators

throttling Operation characterized by the deliberate and/or the infrequent movement of a valve or slide gate to an intermediate position, between the fully open and fully closed, and maintaining the position for periods of time, in response to manually initiated control, for the purpose of changing head loss and/or regulating flow

thrust A linear force on the stem of a valve or slide gate, generally expressed in units of pounds (newtons)

torque A rotary twist or turning movement on the valve stem or shaft generally expressed in foot-pounds (ft-lb) or newton-meters (N-m)

travel stop An adjustable mechanism such as a threaded bolt that is used to limit the operation of an actuator at the fully open or fully closed position

trunnion A pair of posts affixed to opposite sides of the body of a cylinder designed to allow the cylinder to pivot

valve positioner A hydraulic device used to modulate a valve by automatically directing the supply medium to an actuator in response to a position signal

vane actuator A quarter-turn mechanism that uses pressure on either side of a pivoting plate or vane to produce rotary motion and drive a valve stem

This page intentionally blank.

Index

AWWA Manuals

M1, *Principles of Water Rates, Fees, and Charges,* #30001
M2, *Instrumentation and Control,* #30002
M3, *Safety Management for Water Utilities,* #30003
M4, *Water Fluoridation Principles and Practices,* #30004
M5, *Water Utility Management,* #30005
M6, *Water Meters—Selection, Installation, Testing, and Maintenance,* #30006
M7, *Problem Organisms in Water: Identification and Treatment,* #30007
M9, *Concrete Pressure Pipe,* #30009
M11, *Steel Pipe—A Guide for Design and Installation,* #30011
M12, *Simplified Procedures for Water Examination,* #30012
M14, *Backflow Prevention and Cross-Connection Control: Recommended Practices,* #30014
M17, *Installation, Field Testing, and Maintenance of Fire Hydrants,* #30017
M19, *Emergency Planning for Water Utilities,* #30019
M20, *Water Chlorination/Chloramination Practices and Principles,* #30020
M21, *Groundwater,* #30021
M22, *Sizing Water Service Lines and Meters,* #30022
M23, *PVC Pipe—Design and Installation,* #30023
M24, *Planning for the Distribution of Reclaimed Water,* #30024
M25, *Flexible-Membrane Covers and Linings for Potable-Water Reservoirs,* #30025
M27, *External Corrosion Control for Infrastructure Sustainability,* #30027
M28, *Rehabilitation of Water Mains,* #30028
M29, *Water Utility Capital Financing,* #30029
M30, *Precoat Filtration,* #30030
M31, *Distribution System Requirements for Fire Protection,* #30031
M32, *Computer Modeling of Water Distribution Systems,* #30032
M33, *Flowmeters in Water Supply,* #30033
M36, *Water Audits and Loss Control Programs,* #30036
M37, *Operational Control of Coagulation and Filtration Processes,* #30037
M38, *Electrodialysis and Electrodialysis Reversal,* #30038
M41, *Ductile-Iron Pipe and Fittings,* #30041
M42, *Steel Water-Storage Tanks,* #30042
M44, *Distribution Valves: Selection, Installation, Field Testing, and Maintenance,* #30044
M45, *Fiberglass Pipe Design,* #30045
M46, *Reverse Osmosis and Nanofiltration,* #30046
M47, *Capital Project Delivery,* #30047
M48, *Waterborne Pathogens,* #30048
M49, *Butterfly Valves: Torque, Head Loss, and Cavitation Analysis,* #30049
M50, *Water Resources Planning,* #30050
M51, *Air-Release, Air/Vacuum, and Combination Air Valves,* #30051
M52, *Water Conservation Programs—A Planning Manual,* #30052
M53, *Microfiltration and Ultrafiltration Membranes for Drinking Water,* #30053
M54, *Developing Rates for Small Systems,* #30054
M55, *PE Pipe—Design and Installation,* #30055
M56, *Nitrification Prevention and Control in Drinking Water,* #30056
M57, *Algae: Source to Treatment,* #30057
M58, *Internal Corrosion Control in Water Distribution Systems,* #30058
M60, *Drought Preparedness and Response,* #30060
M61, *Desalination of Seawater,* #30061
M63, *Aquifer Storage and Recovery,* #30063
M65, *On-Site Generation of Hypochlorite,* #30065
M66, *Cylinder and Vane Actuators and Controls—Design and Installation,* #30066

This page intentionally blank.

This page intentionally blank.

This page intentionally blank.